COSMOLOGY FOR A NEWBIE

Exploring the Universe: From Galaxies to the Edge of Spacetime

HITEN SHELAR

CONTENTS

ALIENS.

193

THE CONCLUSION

About this book

If

1) You want to understand of our universe and spacetime in an imaginative approach

2) You want to understand deeper about black holes and gravity

3) You are interested to learn various interesting relativity paradoxes and events.

4) You who want to learn more about cosmology

5) You want to know the relation between consciousness and our universe

6) You are a science fiction lover

Then this book best fits you

Preface

Our universe is bizarre. But the world one can see around is not as complicated. We call whatever is around us right now as our Lorentz frames and our bizarre universe is in fact formed from the combination of these normal-looking individual Lorentz frames. The most bizarre object in this universe , 'A black hole', is famous for breaking even these normal-looking Lorentz frames.

Every expert was always a beginner or a newbie in anything he has an expertise in. And as a beginner, one always tries to look at things and explain them, however bizarre they are, by our experience. But when we look at this universe from the cosmic window (our sky), we look at the stuff that is far more bizarre to have experienced in our neighborhood. So, what one needs are the general rules one needs to explain this universe. Our experience being just a small part of these general rules which explain our universe, and we get baffled when it does not work to the universe outside. This book will expand the reader's vision to perceive bizarre nature of this universe and make him understand, what exactly is going on in the fabric of reality that defines our cosmos.

The end of the Earth will not be the
end of humanity

-*Interstellar 2014*

OUR UNIVERSE

Our universe is beautifully designed not only by its constituent but also by the way that every time we try to know about it deeper, it never fails to astound us. The stars we see in the night sky, the air we breathe, the food we eat including our body is made of atoms no one knows which part they emerged from.

Life is the creation of mother earth which was nothing more than a dust cloud before our sun was formed. No sun, no planets, meteors just a dust cloud. What brought every particle of this dust cloud together to form our solar system are the four fundamental forces of our universe which are Electromagnetic force, strong force, weak force, and gravity. The fundamental force of gravity is the weakest among these forces. But still, the force of gravity has a major role in making this universe a possible place. It's because it has a longer range than the other three forces.

All right, that was part of physics, but what is physics? Is our universe governed by the rules of physics? One should remember that physics is only the language that explains various stuff and phenomenon in this universe, so this universe has nothing to do with physics. Physics as a language to describe the universe has failed many times to explain many things which keep happening in this universe. So physics entirely as a subject is still not complete and we just need a theory of everything which is the center of attention of many scientists in this modern era.

In today's busy world, many people just forget that there is something that lies beyond earth. Many people do not even know that there is a subject called astrophysics that studies our universe. Believe me, the more you tend to know about our universe the more you will realize how much you were unknown about the reality you now live in. This new world is no less strange than a science fiction movie. In fact, what makes these things stranger is people's mere ignorance of looking at something outside of the human race. Scientists being human somehow figure this world out and get so much interest in this world which never fails to astound anyone.

THE SPACETIME

The study of our universe is not just the study the what is beyond the earth or outer space. It's the study of everything.

We always believe that something is made of something and that's how we believe that something exists. We know that every object is made of atoms which are the composition of protons, neutrons, electrons, and so on. And this is how reality makes sense to us. If I tell you to imagine any physical entity to be having no subparts then is that possible for you to imagine? (Try it) No one can do that! It's our very basic approach to figure out that something is always made of something and repeat. So what is our universe made of?

For now, let me ask you where you are. And you would give your location. Of course, your address is a name given by a human to a particular place. But have you ever tried to think that where you are 'exactly' in this universe and what are you doing here and what is this universe itself in the first place? Well, lets us keep in mind that we are in space. One should not imagine just outer space now, in physics space means space or let's say a length, an area, a volume, or, a hypervolume.

Imagine for our ease every point in this space being described by a set of numbers. If you are trying to locate a point on a line, we would just use a number. If you are locating a point on the plane, we would use two numbers to locate the point and three for 3-dimensional things, and so on. We are quite familiar with this math at our schools. Let's now imagine 'time'. Hey, have you ever thought about imagining time? Ok from now we will assume it same as a space coordinate. Yes ! a spatial coordinate. The world around you which you can see is all 3 dimensional and when I ask about your position you give me your location in these 3 dimensions. But let's not forget that for any event we do not just describe the location of the event but also the time. For example, you have your train then for you to not miss the train you need to know where that train is approaching and obviously at what time. So let's treat time as a separate dimension, to describe any event successfully. Let's be clear that wherever you look around the universe you look at the past, even if you are describing anything in space you are describing its past! You are describing an event. The description of this event can only be given completely when you speak out its location in space and time.

This space and time Is not separate thing at all as I said one always witnesses and talks about any event in space and its past and so both are not separate at all! So we will not call ' space and time' but rather 'spacetime'. And this is what our universe is made of. The continuum of spacetime is what every constituent of this universe is based on, including the four fundamental forces of our universe.

So 'spacetime' (the building blocks of our universe) is made of 3 spatial which we can see and feel and one time dimension that can't be seen but just felt, making it a 4-dimensional entity. So yes, our universe is four-dimensional This idea of treating time as a separate dimension and this universe was brought by Einstein initially and that has revolutionized astrophysics and scientists' way to look at this universe.

THE GRIP OVER SPACETIME AND GRAVITY

Well, for anything to work smoothly, it needs a set of rules. Yes to have a grip over humans we have government. In subsections like in the office, we have several sets of rules an employee should follow and so to have a grip over them, for kids we have schools which again have several sets of rules. Does our universe which is made of spacetime have the same grip over its inhabitant including every atom to the ray of light and so the rules to follow? Yes of course!

The rules of society or the rules of humans have nothing to do with spacetime. But the fact is that our universe does possess rules for its inhabitants to follow. We at the local scale not even barely go near to violating these rules. We and every object in this universe and this universe which is entirely made of the spacetime continuum are connected by a grip. I would call this grip the 'universal constraints'. Why constraints? Because as one tries to pressurize this grip, the grip eventually becomes a constraint.

The very first thing spacetime is in direct contact with is the ' Mass-energy'. Yes, no other parameter connects spacetime more directly than this. It can be thought of as an ultimate connection between anything existing in this universe and the universe itself. All other parameters are based on the parameter of 'Mass-energy' and it's what gives meaning to every other parameter in physics.

Mass energy and spacetime both have a grip on each other. Let's have a ball freely moving on this 2-dimensional sheet. Now put the heavier ball anywhere on the rubber sheet and now watch the trajectory of the lighter ball. One can perform this experiment at home. What one can watch through this experiment is how the trajectory of the ball changes when a heavier mass is inserted over a rubber sheet. Now Imagine suppressing the 4-dimensional fabric of spacetime to a plane 2-dimensional rubber sheet and imagine the earth as a heavier ball and the moon to be a lighter one! While in the earth-moon case, Moon won't crash into the earth because of zero friction in space contrary to the rubber sheet ball experiment. What I wanted to explain here from this view is that mass-energy curves the spacetime and

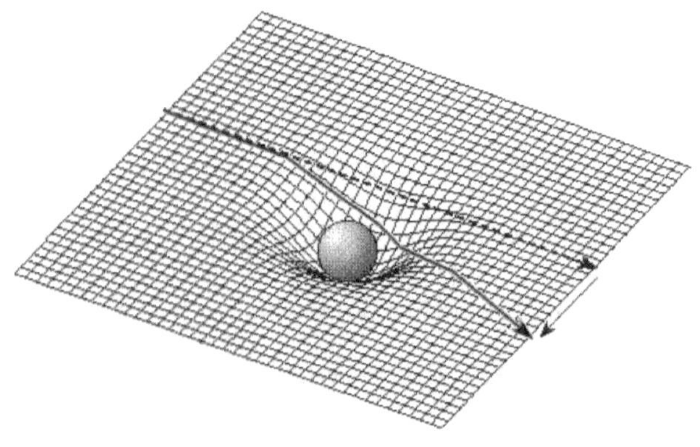

Mass tells spacetime how to curve, spacetime curvature tells mass how to move.

spacetime curvature tells mass how to move. This is the grip I am talking about. Although the rubber sheet example would be a bad idea to illustrate what is going on at a 4-dimensional scale.

"MASS-ENERGY TELLS SPACETIME HOW TO CURVE AND SPACETIME TELLS MASS-ENERGY HOW TO MOVE"

I suggest the reader remember the above sentence as it will be needed throughout the book.

Well, mass-energy tells spacetime 'HOW MUCH' to curve but not how to curve. So what influences the spacetime to curve in a particular way is the density of mass-energy or more specifically 'The Mass-energy density'. Let's treat Mass-energy density as the parameter which tells spacetime "how much and how to curve". The problem is that the density of mass-energy is just one of many factors that tell spacetime how to curve and there are many other factors as well like a frame-dragging effect of the mass which gets some swirling spacetime curvature. So in general so many effects act on spacetime curvature and tell it how to curve and each of them has a different nature. To encapsulate all these effects telling spacetime how to curve, we have a "stress-energy tensor quantity" (don't be afraid, I am not going Into detail). This quantity covers every mechanism from telling spacetime how much to curve and how to curve considering the multiple effects.

From the rubber sheet experiment which was not so accurate, we learned how the moon moves around the earth. But wait damn! We do call it gravity! Yes, spacetime curvature is in fact gravity. Newton was wrong about explaining gravity as a line of force joining the center of the earth and the moon and every other body. It's rather spacetime curvature the stress-energy density created by the earth in the locality of the moon which tells the moon how to move around it. So it's the spacetime curvature, not gravity. So gravity as force loses its meaning. So let's keep in mind that gravity is not a force. The new theory of gravity explained here is the general theory of relativity.

SPACETIME GRIPS BECOMING THE UNIVERSAL CONSTRAINTS.

Mass energy tells spacetime how to curve, spacetime curvature tells mass how to move. This grip is what has made the world around us a possible place and this is how a planetary system and two or more mass pair gravity intersection works.

This grip but has a limit. Yes, there is a limit to how much stress any mass-mass energy over spacetime. These are not only the limits but are the universal constraints that apply to every system of the universe. What if these universal constraints get violated? Well, then the simple answer would be that the fabric of spacetime breaks down. Has it ever happened?

I am talking about black holes. This is the spacetime breakdown caused by the excessive mass-energy density of a star which made it violate the universal constraints of this universe. It's famously known as a 'Black hole'. Another constraint to discuss is the universal speed limit. Yes, there is the fastest speed at which you can go this speed being the speed of light. The closer you go to the speed of light, the more your mass increases and so it requires more and more energy to push the cart to the speed limit. So one requires infinite energy to push the cart to the speed of light and that adds so much mass-energy to the cart that spacetime for it just breaks down.

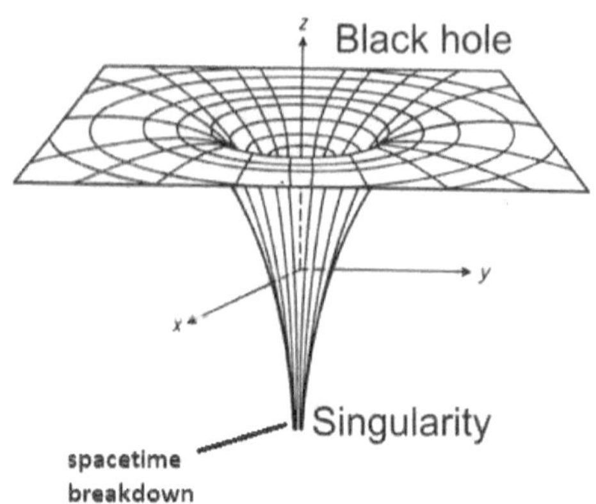

These are just two of the many universal constraints of the universe which apply to every system in this universe. Throughout this book, we will talk about two reference frames. First is the inertial reference frame, which is the frame where an object has zero velocity.

The other is the non-inertial frame where an object has a non-zero velocity. Well, there is nothing truly called a universal reference frame. For example on earth even if we may be at rest but at the galactic scale we are still moving. We use the terms inertial and non-inertial reference frame throughout this book to compare two motions in one's locality and not at the extended scale.

These universal constraints bring out the relativity in the events. Yes, the parameters like time, length, mass, etc are relative. In special relativity, more the speed one has relative to the fixed inertial frame the more one will disagree on the parameters of time, mass, and length with the fixed inertial frame.

Well. Why do we have these constraints in the first place? The simple answer would be that this is how everything in this universe is defined. These universal constraints are just grip for keeping anything in this universe existing in it. Everything thing in this universe has the meaning of its existence deeply associated with the fabric of spacetime. And having a spacetime breakdown in case anything just makes these things non-existing.

The spacetime continuum which our universe is entirely made of has a grip over every constituent of it. This grip has made gravity and other fundamental forces a possible thing. These fundamental forces are the basic tools that created our universe in the way it is. So somehow this grip gives everything in this universe the freedom to get into the non-inertial frame of reference (when forces interact and things accelerate) but the same grip is even having a boundary or a limit for how much extent to have this freedom. This extent is what we call the 'Universal Constraints'

THE LAWS OF PHYSICS

Wherever you go in this universe. In whatever frame, however bizarre it be, the laws of physics would be all constant for you. This is the principle of relativity. And it is followed by everything in this universe. So no matter if you and your inertial friend be strangely disagreeing on one event, the passage of time, your mass, and, your length are gonna be the same in your frame of reference. Let it be a tin paradox where two twins mismatch with age after a time dilation due to high-speed space travel, no twin would watch their reference clocks running faster. So one always carries this local frame bubble with him, everywhere and anywhere in this universe.

Again, how matter how fast you travel you will see the light moving at the same speed difference relative to you (in a vacuum strictly). This is the law of constancy of the speed of light of special relativity. There is nothing special in the light here. What is special here is the speed at which a light ray travels in the vacuum. Yes, light ray travels at the maximum speed limit of this universe. And why light travels at this speed is part of another chapter.

Throughout the process, one might not see any difference in their local frame bubbles but it's the end of the trip when we got to know that we were in some other world (literally!) carrying our local frame bubble with us. And I guess that never fails to astound anyone.

Well however established the laws of physics be, they have their meaning attached to the fabric of spacetime. And the laws of physics would just fail when there is no spacetime (at spacetime breakdowns). One should keep this in mind. The same applies to the universal constraints, however compelling they might be, they can all be violated if they are violated by the universe itself. This all things would be given in detail later in this book.

I am letting readers know about some of the content required to get a better understanding of later super interesting chapters through some of these initial chapters to get some basic knowledge and the journey ahead would be full of interest if one has a grip over these initial chapters.

THE UNIVERSE AND PHYSICS

The universe did not always exist, it has its own birthday. Our universe was born from an infinitely small point to what it is now. We call this birth process of this universe ' The Big Bang '.

Physics > Chemistry > Biology

The three subdivisions of science which are Physics, Chemistry, and Biology have not to do anything with our universe. These subjects are just humanity's attempts that best describe the universe under our vision. Just after the big bang, we had the birth of physics. TThesephysics made the possible interaction between various particles and waves in this universe. It joined the various subatomic particles to form an atom and hydrogen was formed. The fusion of different atoms created a set variety of different elements and other subatomic particles and gravitational interaction made these atoms come together to form a star and its fusion process to maintain its existence and power our planet. The interaction of the variety of these atoms has given rise to chemistry. The complex chemistry gave rise to bio-molecules and we had biology. Biology gave rise to us, questioning how this universe works. So all the things around us including chemistry and biology are what physics has given a rise to (generally speaking). And physics has its meaning entirely associated with the continuum of spacetime.

Maths which might be the reason many don't like physics is beautiful. To be honest, many don't like it because society tends to compete and not focus on the beauty of math. An excellent teacher will always try to show this beauty. Even it was once my nightmare. But later in life, I was gifted with one of the great teachers I ever had, who taught me the beauty of this subject. The thing is that it describes reality so well. It can nearly describe everything in this universe. It's the language of our universe, it applied in physics it makes us see things where our vision is far from reaching, helps us predict the motion of the stars, helps us to encapsulate the laws of physics that are globally applicable and so it saves a lot of our time as well. It gives a precise measurement to achieve for our missions to accomplish. But again as a subject, it's just our attempt to explain this universe, and the universe has nothing to do with maths. But as a subject, yes it's beautiful to apply it in physics and it greatly predicts the results.

HAUNTING GRAVITY

We know, if a force is applied to anything, it will move. But we also know that for any object to move we need a force. I mean an object moving without any applied force in real life would be ghostly. Well, the same thing is happening with us and everything around us now itself. Yes, the objects are moving downward without any applied force and that happens on the entire globe. Isn't that haunting? No. because no one is gonna get afraid of ghosts if the same ghost meets you daily and once you get a habit of it, it is not ghostly at all. If we were born in an imaginary world where gravity never works and then suddenly we made access to gravity, yes then gravity would be so much ghostly. But now in the real world, gravity is our daily life ghost force. This same force has made our universe a possible thing, including our world. We are so much prone to this force. The same thing again for the electromagnetic force, we know two magnets attract but we don't find it ghostly. The rest of the two forces work the same way, but we can't see them interacting at our scale.

For a physicist, there is nothing ghostly in gravity. Newton said the force of gravity is a line of force that connect every mass in this universe and gifted the world, the equations to reach outer space. Locally and non-relativistic (speed not comparable to the speed of light), newton's laws work quite well. Einstein's description of gravity is beautiful, it's great, it's my favorite, and it describes gravity so well. According to einstein's general theory of relativity, gravity is the curvature in spacetime. And when the spacetime that defines us get affected by this curvature, we feel some acceleration, this is the acceleration of gravity. Everything that has mass energy possesses a spacetime curvature, including the tiniest subatomic particles.

Gravity is the weakest of all the four fundamental forces of the universe. So at our scales, the spacetime curvature is negligible. But provided we have a lot of mass-energy in this universe, this weakest force has the strongest gravity force influencers (black holes!) in this universe.

Thanks to Einstein's gravity we could explain the trajectories of the planets which newton's gravity failed. For example the perihelion of mercury. The force of gravity is just not the game of mass but also has a role of energy it possesses. Mass energy just tells spacetime how much to curve. But in reality, we need the stress-energy tensor (page no.10) to describe the spacetime curvature possessed by any mass. Newton's gravity never took stress energy into the account. This is the reason it fails to explain the trajectories of many systems in this universe. But yes at local scales and non-relativistic speeds, they do work extremely well.

The fabric of spacetime is the fabric of reality. It is something on which we have our existence in this universe assigned, we can feel space and time around us (not forgetting the fact that it's 'spacetime' and not 'space and time'). And when Einstein proved that gravity which is so much known to us is the curvature in the fabric of reality, then what theory is more astounding than that? (literally!).

DON'T UNDERESTIMATE GRAVITY.

Gravity being the weakest among all fundamental forces has the highest number of force-exhibiting agents. Particular mass energy always carries spacetime curvature with itself (so the gravity) if not other forces. So one can't find any mass energy that does possess electromagnetic, strong, and weak nuclear force but not gravity (strictly) and no vice versa. Somehow the weakest force of this universe has been gifted with the highest number of ingredients that me this force a possible thing. The force of gravity also has the longest range of than other three fundamental forces. So eventually every mass is going to attract every other (if provided enough time).

The abundance of mass-energy in our universe has given rise to gigantic structures like stars, and black holes. Galaxies, superclusters, etc in our universe. Black holes are among my favorite among them. Their size range from the size of an atom to a thousand time larger than our solar system. What makes these guys very famous in astrophysics is their strong gravity from where no light and not even information is supposed to escape. I prefer calling black holes a beast because of their appetite. Their normal snack contains stars that are many times larger than our sun and while doing so they continue to become larger.

The most famous black hole is the stellar black hole. The process of their formation would be discussed in the next chapter. Black holes are the strongest gravity influencers in the universe, in fact, no force influencer of whichever force is stronger than the black holes. So the weakest force of the universe has the strongest force influencer in this universe (Let's keep this in mind).

The stellar-sized black holes are normal-sized. The gigantic, supermassive, and ultramassive black holes are the ones which are located in the center of the galaxy. Quasars, which are one of the brightest objects in our universe are powered by these black holes. The next chapter is the recipe for a stellar black hole.

RECIPE OF THE BLACK HOLES.

Let's say we have a star that has a mass between two to three solar masses. In every star, the core is always denser than the outer layers of the stars. This core having energy density always tries to become denser and denser due to the still inner layers of the star and so it always keeps pulling the star layers inwards. But for a star to maintain its status of a star and maintain its size and outer layers, it has the counterbalancing force against the inward pulling gravity. This equilibrium which maintains a star's density and makes a star a star is called the hydrostatic equilibrium.

The energy from this hydrostatic equilibrium is derived from nowhere but from the strong force. Yes, the same force which is responsible for binding a proton with a neutron. A newborn star begins by fusing hydrogen to helium until there is not enough hydrogen to maintain the star and then helium starts fusing to carbon, after which carbon fuses with helium to form oxygen, oxygen plus helium produces neon, neon plus helium produces magnesium, magnesium plus helium gives silicon, silicon plus helium gives sulfur, sulfur plus helium gives argon, argon plus helium produces calcium, calcium plus helium gives titanium, titanium plus helium gives chromium and at last chromium plus helium produces the iron. Enough! in this battle between the strong force and gravity let's accept that strong force has a limited fuel supply in the battle. Because iron is just unfusible. But the stellar mass is always present to make the star more than ever before. This turns the star into an object called a white dwarf, which maintains its layers by quantum mechanical pressure (will be discussed later) called the 'electron degeneracy pressure'. Let's be serious now, this pressure has enough potential to counterbalance the inward pull of gravity of the stellar mass for eternity. And it does not run out of fuel, as it does not run on fuel! The electron degeneracy pressure is not due to any underlining new force. It's just nature's way to avoid the violation of physics.

But we are dealing with the recipe of a black hole, so we are ready with enough mass to make degeneracy pressure fail as well. That's the reason why I took a star with a mass of 2 to 3 solar masses heavy as the more the mass, the stronger the spacetime curvature or the pull of gravity is. A star less massive than we assumed would just create a white dwarf and nothing more than that. Our sun is expected to have the same fate.

Let's get back to our ingredient star with a mass of 2 to 3 times the mass of the sun, having gravity stronger enough to break even the electron degeneracy pressure. The main ingredients of the white dwarf which provide an electron degeneracy pressure are mainly basically electron protons and neutrons. Having our pull of gravity enough, the electron degeneracy pressure fails and the electron and proton in the white dwarf are forced to combine to form a neutron and a neutrino particle. And then we have a thing called a 'Neutron star' with an upcoming BOOM!

These booms are the most violent events ever recorded. This boom is called a supernova. This boom is caused by shading the upper layers of stars very violently. But what causes that bump? This bump is caused by a ghost particle that travels like a ghost without interacting with any mass. I am talking about a Neutrino particle. This particle is continuously thrown out of the sun and trillions of them pass through us and everything around us like a ghost. These particles interact with a weak force (which is stronger than gravity!). But keeping in mind that the outer layer of the star is so dense, there remains a good chance of very few out of a very very high number of neutrinos to interact with the stellar upper layers and blow it in form of a supernova. Which is the most violent event in the universe.

So the most violent event in this universe is caused by the least interactive particle in this universe. Isn't that beautiful?

Back to the recipe of the black hole again. So now we have a neutron star with us followed by a boom which we just had a long discussion on. The neutron star again has its quantum degeneracy pressure (will be discussed later). Again it is stronger enough to keep the neutron star stable the eternity. But as we are prepared with mass high enough (2 to 3 solar masses), we still break throughout this quantum mechanical pressure. Now we have our beast, a black hole! A place inside which every law of physics fails including the general relativity black hole is a breakthrough through spacetime. Our laws of physics simply won't work out as their basis which is assigned to the fabric of spacetime is shattered over here.

A STAR AND A BLACK HOLE

In the last chapter, we discussed that simply having mass-energy is not going to create a black hole instantaneously! Somehow the star could turn into a black hole without even forming a star, what we needed was just pure mass-energy. But in reality, the majority of mass-carrying particles in this universe do possess other properties as well and so the other two of three fundamental forces come into account. These same forces forces forces maintain the star layers in our ingredient star in the last chapter before it turns into a black hole. In absence of these three fundamental forces, gravity would just have formed black holes, and nothing else.

But the star and a black hole formed from it although having the same mas are very different from each other. Black holes' gravity is much more violent. So much violence that the fabric of spacetime breakdown in its center. So what makes a star a 'star' and a black hole a 'black hole' is the mass-energy distribution. A black hole is very much denser than a star. Both have the same mass, but what keeps a star's density less is its hydrostatic equilibrium for which the strong is responsible. So it's not that we have bulk atoms and we get them together and instantaneously we have a black hole. We have to wait for the stellar fuel to end and again cross two degeneracy pressure barriers to break the fabric of spacetime and form a black hole. But eventually, a star 2 to 3 times massive is going to form a black hole. And this time between a star to black hole transition is what gives chance for life to arise on earth, giving birth to our world.

The fusion energy of a star, derived from the strong force which has no direct relation with spacetime, whatsoever is maintaining the spacetime curvature of a star from breaking into a black hole. This is amazing to know as it's a strong force that is controlling spacetime indirectly although temporarily.

VIOLENT AND NON-VIOLENT BLACK HOLES

Black holes are the strongest gravity source in the universe. These beasts can be further classified into violent and non-violent ones. The brutality of the black holes speaks a lot of past about the stars which they had formed from. Except when they collide with other black holes which kind of mess up those imprints. These imprints are left during the star-to-black hole transformation process, which is very violent. Let's say a star with 10 solar masses will transform into a black hole faster than a star that has a mass of 4 solar masses. So we can say that the transformation process was violent in the case of massive stars and this brutality gets imprinted on the fabric of spacetime and the fabric of spacetime gets more violently dragged, creating a less continuous flow of spacetime into the center of the black hole. The more continuous the flow of spacetime is, the less violent the black hole. So in the case of stellar black holes, black holes formed from a massive star is always more violent than a less massive one. This applies to only this case although. In the case of supermassive black holes which sit at the center of many galaxies including ours, the spacetime curvature is less violent and we have a continuous flow of spacetime to the point where spacetime ceases to exist. Brutality in the case of spacetime curvature means how strongly a curved spacetime curvature departs from a flat one. The stronger this happens the more violent the black hole is. Massive stars once done with the fusion process attain transform into a black hole faster due to huge mass and so the more gravitational pull. The faster this happens, the stronger the flat spacetime around the star departs from flat to in moving curved spacetime and so the more violent it is. If the star is rotating around its axis the then same is imprinted on the spacetime in form of a frame-dragging effect. In the next chapter, we will see how it is to voyage into violent and non-violent black holes.

WHAT IF I FALL INTO A BLACK HOLE?

Well, what's gonna happen depends on which black hole you choose to fall into, violent or non-violent. Let's go with violent black holes first. Let's say you are voyaging near this black hole and you find your way to its accretion disk and are revolving around it at very high speeds. As time passes by you will go near and near the black hole. Let me make you know that you have limited time to boost your rocket off, to escape from the black hole. As you near a black hole, the escape velocity for you keeps on increasing until it attains the value of the speed of light on the event horizon. When this happens, escaping a black hole would be like violating the laws of physics. But again let's not forget that this black hole Is a violent one, so the voyage is not going to be as smooth as butter at all.

In the case of violent black holes, the spacetime curvature departs very strongly from the nearby flat universe. So strongly that, your head and legs go in two different Lorentz frames (if you are heading toward a black hole vertically). And this makes you feel a stretching tidal gravity. Yes, the same tidal force by which the moon gets sea waves on earth and by which the moon is tidally locked with the earth. In the case of the earth the and moon, this tidal gravity is only apparent if two points of the body suffering tidal gravity are separated by a distance high enough. This tidal gravity is due to differences in spacetime curvature at points of the body separated high enough. Due to this, the near-most part of the moon is more under the influence of the earth than the other, the combination of this effect and the centrifugal force on the moon ultimately then has the stretching effect over it, making its one side getting tidally locked with the earth.

Near enough, to our violent black hole, we have high centrifugal force over us and stretching tidal gravity every micrometer. This tidal force will separate every atom of your body before you make it to the horizon. This process is called spaghettification. So voyaging into a violent would not be a great idea at all!

Let's now voyage around a supermassive non-violent black hole. Again, you have no chance to escape this black hole if you reach the horizon where the escape velocity is the speed of light, and escaping becomes equivalent to violating the laws of physics. In the case of non-violent black holes, curved spacetime departs from the surrounding universe's flat one smoothly (at least much smoother than the violent black holes). So there is strong tidal gravity but no spaghettification and you can continue your voyage in your spacesuit still further. Once you cross the event horizon you start heading towards singularity. When you reach the singularity, this universe will deduct your mass from you! (read next chapter) and then your body has no existence. But why? Because there is no spacetime after that! We have our meaning deeply assigned to the fabric of spacetime and no spacetime would just mean no existence. So unless there are bulk beings to save you like in interstellar by blocking you inside a tesseract, the end is fixed inside a black hole or not? Well, that would be the part discussed later in this book

THE BEAUTY OF PHYSICS

Physics is beautiful and it successfully explains our universe to a great extent if not entirely. For example, the law of conservation of energy says that the total amount of energy in this universe is constant. And as mass is energy, we have a corresponding law of conservation of mass. Let's say a particular thing after falling inside a black hole loses its existence in this universe. So does that include mass itself? Does the total amount of mass energy of the universe decrease in this case? The answer is no.

The same applies to properties of charge and angular momentum as they have their own law conservation. When a mass falls into a black hole, that simply does not just vanish, otherwise, that would be a violation of one of the basic laws of physics. As soon as any object cuts itself from the entire universe at the singularity, its mass gets surmounted over a black hole. Yes, its mass adds up to the mass of the black hole. And same is the case of charge as well. What surprises me is when we put a body with some angular momentum associated with it into a black hole. It just simply adds up to the black hole's spin around its axis.

It's beautiful to imagine how the very basic laws of physics are conserved here throughout the process. In process of violent stellar collapse into a black hole, there is no violation of physics. Except for some topics, physics works extremely well to describe our universe. The equation $E=mc^2$ is one of my favorite equations in physics. In this equation on the left side, we have energy, and another side we have mass. General people mostly misunderstand mass with weight, they perceive mass as a force and that is wrong. Mass is but somewhat a different concept. We define energy as the ability of any system to do work. And mass-energy equivalence describes mass as energy which is the ability of a system to do work. I mean imagine using no fossil fuel, but mass as energy to push the cart. This is the beauty of this equation. It says that a small amount of mass can be converted into an enormous amount of energy. So mass stores so much energy. And this result is entirely derived from the principles of the special theory of relativity. This mass-energy equation is not what has Einstein discovered, it been found in many solutions in physics. But the credit goes to Einstein because he particularly stated that mass is energy and that's why he truly deserves this credit. Well, to move further in understanding the black hole beasts, I need the reader to know quantum mechanics at a very basic level. So let's go with that in the next chapter.

A BIT OF BASIC QUANTUM MECHANICS.

What idea does the word quantum give you? A quantity or discreteness? Yes! This is what quantum mechanics talk about. This field of the study explains all fundamental forces of the re in terms of the exchange of discrete or quantized energy packets. It explains gravity force as an outcome of an exchange of massless gravitons between two masses., the electromagnetic force is an exchange of photons between two charged particles, gluon particle exchange is responsible for the strong force and W and Z bosons are responsible for the weak force interaction.

Generally speaking, quantum mechanics is the field of study that describe subatomic level physics extremely well. Newton's theories truly just apply to the normal world physics and they just fail at the largest and the tiniest scales of spacetime. Einstein's general relativity covers the physics of astronomical objects and quantum mechanics covers the world of tiny subatomic particles. What makes the world at a subatomic level stranger is the knowledge we lacked until the advent of quantum mechanics. Quantum mechanics is famous for its language of probability. This language won't make sense at our scales at all! If it did then you might cross the wall without crashing over it, you had no definite position and would have kept continuously oscillating in position everywhere. At our scales, we calculate various quantities in physics and believe that they are precise. This preciseness just does not exist at the It's the scale where only the highest probability outcome has a good chance to take place. And so we have 'The most expected outcome' rather than a precise value outcome at the subatomic scale. Finding these probabilities and predicting the future motion is what quantum mechanics does. More precisely, quantum mechanics is the analog of classical or Newtonian physics at the subatomic scale.

The most bizarre part of quantum mechanics is that it describes matter as waves! What? Ycs, I know what I am talking about! So if you throw several electrons on through a slit, they will land on a detector in form of an interference pattern. This same pattern can be found when a light wave is passed through a slit. So, that means the matter is a wave. Several phenomena like the photoelectric effect which yielded Einstein a Nobel prize can only be explained if the light is treated as a particle. Generally speaking, that means the matter is waves and wave is matter.

The masses of the subatomic particle are less, so it takes relatively less energy to boost them to the -to-light speed, which is very common at that scale. Einstein's special theory of relativity and quantum mechanics describes the motion of these particles extremely well.

Well, if the matter is a wave. Then are we the waves? The answer is yes! We as a wave don't show the quantum mechanical behavior and that is because our wave function is sharply spiked to where we are leaving no uncertainty to whatever takes place at the quantum level. The more your mass is, the less wave-like behavior you possess (Keep in mind that every 'matter' is a wave! In reality, the wavelike behavior is only noticeable at the subatomic scale.

Can any quantum mechanical event happen with a normal day object provided we wait long enough? Yes, it can happen. But we have to wait for the entire lifetime

of the universe for any such event to happen. Now the reader has a bit of knowledge of quantum mechanics, let's now proceed to see what it says about the singularity of a black hole.

IS SINGULARITY REAL?

Our universe is entirely made of a spacetime continuum. And reality has its meaning attached to it. When an object meets the singularity or spacetime breakdowns its mass, charge, and angular momentum gets deducted from it and gets added to the black hole.

If I ask you to let's say destroy a ball. You might put it into the hydraulic press, melt it, and shred every atom of the ball. But there is nothing better than putting it into spacetime cutoffs to perform this. Since it then separates itself from the entire universe and not even any of its information remains. What's more dangerous than this?

Keeping this theory aside, many physicists believe that the singularity is not a spacetime break-off just a finite tiny volume, so tiny for time to exist over there. Time simply has no meaning at this point. This finite volume is at the scale of the Planck length. Planck length according to quantum mechanics is the smallest possible length in our universe. According to quantum mechanics, the singularity is a point with dimension at the scale of plack length and where time out of spacetime breaks and we just have a space with a tiny Planck scale volume.

The world at this scale is even more bizarre than what happens at the subatomic scale. At this dimension, quantum mechanics says that what exists is a quantum foam. The image below represents the quantum foam.

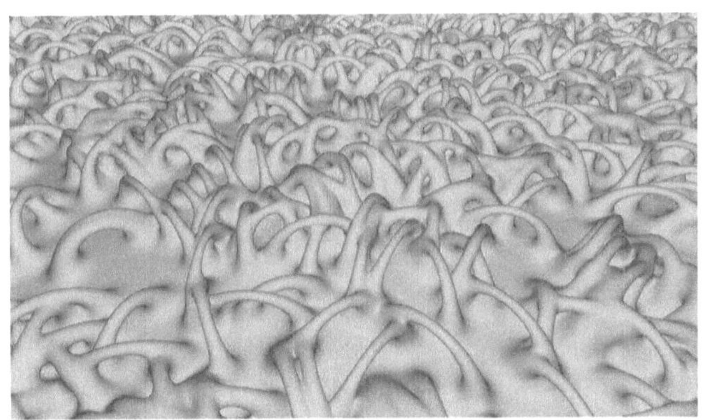

The quantum foam is what spread over the entire universe. And at this scale, a large scale of the creation of virtual particles with any energy can be seen. The bubbles in this image are just the representation of the randomness and freedom of energy these particles have to be created. The most interesting fact is that not only particles but also the tiny universe can be popped out of this quantum foam. Our universe is suspected to create itself the same way.

Remember, when you visualize the quantum foam, just forget about the 'time'. At this scale, 'time' just ceases to exist.

So according to quantum mechanics the 'space' part of 'spacetime' stays alive at the singularity and the part of space is unbreakable because of the limit of quantum mechanics on the unit of length.

Explaining the singularity as a spacetime breakoff is although my favorite theory. Why? We'll cover that later in this book

ARE THE BLACK HOLES IMMORTAL?

Black has extremely long livability but these beasts are permanent in this universe. They lose their mass through a process called Hawking Radiation.

This universe is lined everywhere else with a thing called 'Quantum Foam'. Just imagine, where there is spacetime there is quantum foam. It's located all over the universe. It is a very basic layer of energy that is located at every point in the universe. At this level, something creates itself from nothing and again meets nothing. Got jumbled up? Let's go into detail about that.

At the scale of quantum foam, some new process happens which cant be seen at our scale. Imagine. This process is what defines the quantum foam, which is a very basic energy layer of this universe. Imagine a hypothetical zero energy condition of the universe, where there is no quantum foam. Now imagine you create two particles, one normal matter and antimatter (it's anti-part) in this zero energy empty vacuum. with the same mass and energy, just with an opposite charge. And after a very short interval of time, these virtual particles return to annihilating themselves.

That sounds bizarre. I mean how can we have something come out of nothing like magic? And mainly, isn't it a violation of the conservation of mass law? Well, no! Although this law is violated, it happens just for a very very small intervals of time. So it's just a momentary violation and at the end of the day, the system comes to zero energy again, which is permissible.

Another question is that, can we see normal objects like chairs, spaceships, or anything showing the same behavior? Before answering this question I want to introduce the reader to an equation that predicted the existence of the virtual particles in the first place.

$$\Delta t \Delta E \geq \frac{\hbar}{2}$$

This is Heisenberg's uncertainty principle equation. Here 't' stands for the time, 'E' stands for the energy and 'h' is Planck's constant ($6.62607015 \times 10^{-34}$ m^2 kg / s). Please don't get baffled, I will explain what this equation is trying to convey.

The equation says that it's ok to see small energy changes for very small intervals of time. For a scale larger than the subatomic scale, we have a change in energy very high. So this can happen only for a very very short amount of time. So, can we have a spaceship behaving like virtual particles? The answer is still no. That's because there is a limit to how small a particular time interval be. This is what we call the 'Planck time'. When you calculate the energy associated with the normal day objects, it will come out to be a very large number (at the scale of 10^16 meters). If one substitutes this energy change value in the uncertainty principle, then the amount of time permissible for this object to come into existence comes out to be lesser than the 'planckPlanck, which is the smallest possible time interval possible in this universe. That's why everyday objects can't show such behavior.

Coming to the title of this chapter, imagine the same virtual particle pair creation process happening extremely near to event horizon of a black hole. At a point near the black hole, one has a significant tidal force at every atomic scale distance (the same tidal force which results in spaghettification). This tidal force makes one of the two virtual particles fall inside a black hole and the other escapes. The other particle escapes due to a strong stretching effect of the tidal gravity, making a particle nearer to the black hole fall inwards and another thrown out. This also creates a thing called negative mass which we shall discuss later in this book. The thrown-out particle flies at relativistic speeds and escapes the black hole. This kinetic energy which boosted particles at relativistic speeds got from nowhere but from the mass of the black hole itself! How? well, let's discuss that in the next chapter.

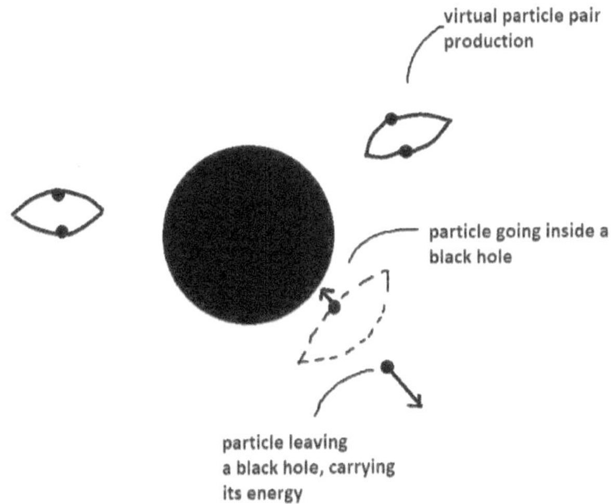

This mass deduction process is very very slow to make the black hole completely get evaporated. By this process alone it might take a time comparable to 10^100 years (one googol!) which is a very very very long period! Normally black holes eat a lot more than what they lost via hawking radiation, so hawking radiation almost has no change for black holes sitting at the center of the galaxy.

This universe again proves that nothing is permanent in this universe, This is how beautifully, this universe is designed.

IS GRAVITY AN INFINITE ENERGY SOURCE?

I hope the reader is familiar with the technology called an orbital maneuver. If not, then still no problem.

Imagine a space shuttle moving near some planet. The orbital maneuver is the idea, which can help the space shuttle to speed up without any extra fuel usage. The idea is to take a moving rocket under the gravitational influence of some planet and take the help of the gravitational acceleration offered by a planet to further increase the speed and once that happens the shuttle will use its thrusters to escape the planet's gravitational pull. This plan is only profitable if the kinetic energy received is more than the energy rocket used to move out of the gravitational influence.

But hey! Although the idea is good, is gravity a free energy source? I mean, from where did we borrow the kinetic energy for our space shuttle? The first law of thermodynamics says that the total energy of this universe remains constant. So of course we are charging something else for acquired kinetic energy. More specifically, we are borrowing it. We are borrowing it from nowhere but from the planet's mass! As mass is the same as energy, we are deducting energy from the planet in the form of mass and using the same energy to boost our space shuttle.

The same thing happens in the case of Hawking radiation when one of the particles falls inside the black hole and another escape it. The escaping particles get their energy from nowhere but from the mass of the black hole itself and this is how black holes evaporate.

Do you want more examples? Ok! The moon is tidally locked with the earth. Without this force, we could be able to see the other side of the moon. So the earth is having a stretching effect over the moon? The energy for this stress to occur is deducted from nowhere but from the earth itself in the form of its mass. Again, what do you think the ocean tides get their kinetic energy from? It's deducted from the mass of the moon. Saturn has stretched out every asteroid into the smaller size ones, forming its beautiful rings. Again, Saturn has given a part of its mass for this to happen. Europa (Jupiter's moon) has a liquid water ocean even after being far away from the sun. A thing that is maintaining Europa's liquid water is nothing but Jupiter's tidal gravity which is obtained from the mass of Jupiter.

Our ocean tides still barely affect the mass of the moon. As the small amount of mass store immense energy, the borrowed energy barely affect the planet's mass. So imagine the gravity or the spacetime curvature, just like an energy offering. Anything that accepts this offering and does some positive work, affects the mass of the source of gravity.

HOW CAN GRAVITY ESCAPE THE BLACK HOLE?

Nothing in this universe can come out of a black hole once felt inside it, not even the information. So how is gravity able to escape the black hole and reach you?

Do you think that the gravity you are feeling now is coming out of the center of the earth? The gravity you are feeling right now has nothing to do with the center of the earth. Gravity or spacetime curvature shapes itself based on mass energy and its distribution. The gravity or spacetime curvature, one now feeling is the measure of spacetime curvature in the locality.

The gravity one feels is due to the spacetime curvature one has in his Lorentz frame. So gravity does not travel from the center of the earth towards you. Does the spacetime curvature at the center affect the spacetime curvature in your Lorentz frame? No! The overall spacetime curvature depends not on parts of it but just on the mass-energy distribution. So no Lorentz frame in spacetime curvature has to do anything with the other. For example, at the center of a black hole, spacetime ceases to exist, but the surrounding spacetime curvature around the event horizon has to do nothing with that.

In the case of a stellar black hole, the spacetime curvature is just an imprint of the star it ever was. So to conclude, gravity does not escape the black hole.

BIZARRE OBJECTS IN SPACE.

Under our observation at the local scale, it seems that this universe has an abundance of planets, stars, asteroids, and comets. Nowadays we are so much familiar with them that only astronomy-crazy people look up at the sky to wonder about them. But there is something out there still more bizarre. For example, a white dwarf. We imagine this object as a normal white star-like-looking body. But the fact is that its stranger than that. Stranger because it's not made of material things like any of the stars, planets, asteroids, or anything else. It not even can be called to be made of atoms. The same applies to neutron stars as well. They are just simply made up of neutrons. They are so dense that a spoonful of neutron star matter would weigh as heavier as mount Everest.

Have you ever heard about quark star? It's a hypothetical star that is suspected to exist in between the neutron star to black hole transformation. Imagine an astronomical body completely made out of quarks. What about the black holes? These objects cant thought to be real at all. That's because the fabric of spacetime on which reality is dependent is broken inside the center of a black hole. So black hole is not a planetary or any stellar object. Although it behaves like other astronomical objects, finally it proves that is not what it seems.

Have you ever wondered how planetary strikes would be different from the collision between a black hole and a planetary object? We have many craters which give us a glimpse of how asteroid strikes. But the collision of a black hole with our planet just has a different story.

Black holes are mostly famous for their ability to swallow things. Its collision with some other planetary or stellar object is what no one talks about, that is because the planet would be spaghettified the moments before it makes contact with a black hole. But that happens in the case of stellar or massive black holes which have a huge domain that extends far outside the event horizon. So in our example, we are having a black hole that is roughly equal to the size of a hydrogen atom, so it has no larger domains to shred every atom of the earth.

We will assume here that the black hole in our example already has its accretion disk built already. The accretion disk of this black hole would be immensely hot. This accretion disk is enough luminous to shine like 100 Hiroshimas. That's so bright for an atom-sized black hole. But the same luminosity is even important for a black hole to maintain its accretion disk.

Yes, the accretion disk of a black hole is maintained because of the radiation pressure of the accretion disk. This radiation pressure counter-balances the black hole's inward pull, maintaining the stable orbit of matter around it and so forming a stable and hot accretion disk. This accretion disk is responsible for the violent collision of a black hole with our planet.

Imagine something brighter like 100 Hiroshima entering our atmosphere, heading towards the land. When it reaches very near to the surface of the earth the radiation pressure of the accretion disk will firstly dig the surface of the earth. The matter from the dug part adds up to the accretion disk of a black hole. Once that happens, the black hole swallows the part of the preexisting accretion disk to welcome the new material, I mean that's for keeping the accretion disk stable. Every time a black hole digs the surface throughout the process it keeps swallowing old material to welcome the new one. This makes a black hole not stop in between like an asteroid to create and crater and that's it. Rather black holes kind of penetrate through the entire planet creating a global earthquake. So the asteroid collision just ends up creating a crater on the planet, but in the case of a black hole, we have a very deep hole than a wide crater.

That is how a planetary collision is different from a collision of a planetary object with a back hole.

BLACK HOLES AND THEIR GRIP OVER SPACETIME.

Remember, a black hole is just not a hole in space, it's a hole in spacetime. So it has so many things to do with time as well. Time is the component of spacetime that makes spacetime flow. When one is rest in space, time completely moves on, and when one travels at the speed of light or very near to it, time elapsing stops. This stop in the passage also occurs when one nears a black hole.

As many might imagine, a black hole is a hole in spacetime from which nothing can escape, not even light, and so on. Common, that has now become an old story. It's now a time to talk about something that a black hole influences and we can't see it but feel it. I am talking about black holes' influence on time. It would be nuts to talk just about time and no spacetime as both are attached and gives meaning to our universe. So let's see what a black hole does to the fabric of spacetime and we shall see in this chapter, how this effect on spacetime can be felt when we feel time.

To be honest, I kind of don't believe in astrology or future predictions. But when we talk about black holes, these holes in the fabric of reality do govern our future. Again, I know what I am talking about.

The spacetime does flow. We feel the gravitational acceleration. that is because spacetime tends to flow. Even if we might rest in space, the spacetime that defines us does flow. This moving effect of spacetime is due to flowing time. That's why we may conclude that gravity can't exist without time. It's a universal law that spacetime tends to flow.

For now, I want the reader to suppress 2 dimensions from space, for our ease to imagine. And now we just have one dimension of space and one time in our spacetime to imagine. Imagine this flowing spacetime into a black hole the same way the water makes it into the washbasin. In our spacetime diagram, we are somewhere in the middle, which we assume to represent our present. What below our present is the past and above it is the future. The future condition of our spacetime is what going to be our present. For example, if a future worldline has some curvature due to some other massive planet, then that would somewhen reach us in the form of gravitational acceleration. And after that, it goes into the past. The trajectory through this spacetime is called our worldline (which in reality is four-dimensional).

A black hole is a hole in four-dimensional spacetime and it not only swallows the space but also time! Imagine our world line into a black hole like a river. The part first enters is the front side which is the future part (just because time runs from past to future in our universe).

Once our worldline goes into a black hole, it has no chance to escape it as could and according to the laws of physics, nothing can. Assuming your future worldline is flowing into a black hole. Your present has to follow the same footsteps automatically due to the flowing nature of spacetime. In other words, the track of the future world line can change by boosting off our rockets. But that simply won't work out when you are inside a black hole. This means that a black hole governs your future. As once your future world line is inside a black hole, your present is destined to go inside it.

That means a black hole governs the future. Do not relate this chapter to your life goals, please. This entire chapter was just about the fabric of spacetime. And remember our destiny is in our own hands, and we are the only driver of it.

CREATING A SO-CALLED BLACK HOLE-LIKE HORIZON.

What makes a black hole separate from this universe is the fact that no event inside a black hole can be seen from the outside universe and no event from the outside universe can be seen from inside the black hole. The boundary from which this cut-off of the events occurs in the case of a black hole is called the event horizon. This topic shares an idea that can let us create a horizon around ourselves, if not a black hole.

So as one knows that a horizon does the work of separating events, the idea is to create a virtual horizon around yourself that would separate you from that event. Take into the picture, the same spacetime figure we imagined in the last chapter. Now this time we are extending this spacetime to the entire universe. I know that would be very huge, so to make a diagram fit in our imagination. So, let's perform some coordinate transformations.

The coordinate under our imagination has one dimension of space (represented as x-axis) and one time (represented as y-axis) :

Now, let's draw a distance-time graph of light on this coordinate plane :

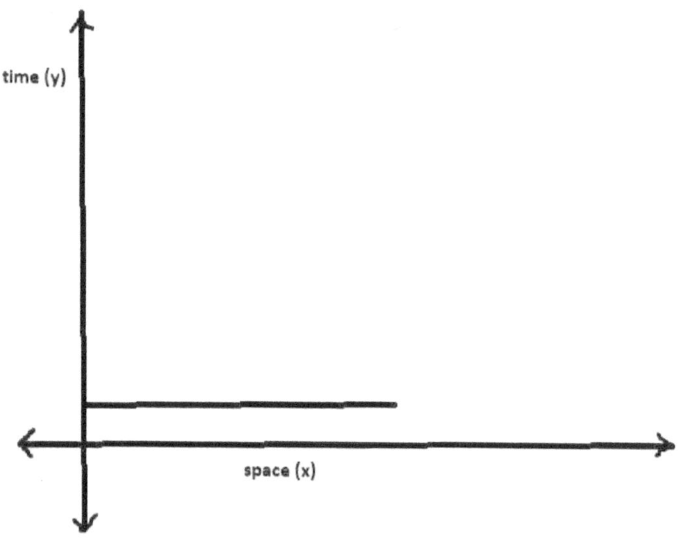

Time does not elapse for light, so we describe the distance-time graph for light as a line parallel to the space axis.

Now, to imagine the entire universe, we perform the coordinate transformation. Imagine our universe like a chocolate wrapper, now by tying it up on four sides we are avoiding imagining the infinities. Imagine we as an observer as chocolate inside the wrapper, which we shall deal with after a few illustrations. The wrapping of this universe is done in a way in a way that the distance-time graph of the light mentioned above becomes tilts by 45 degrees. And this is the spacetime graph we get :

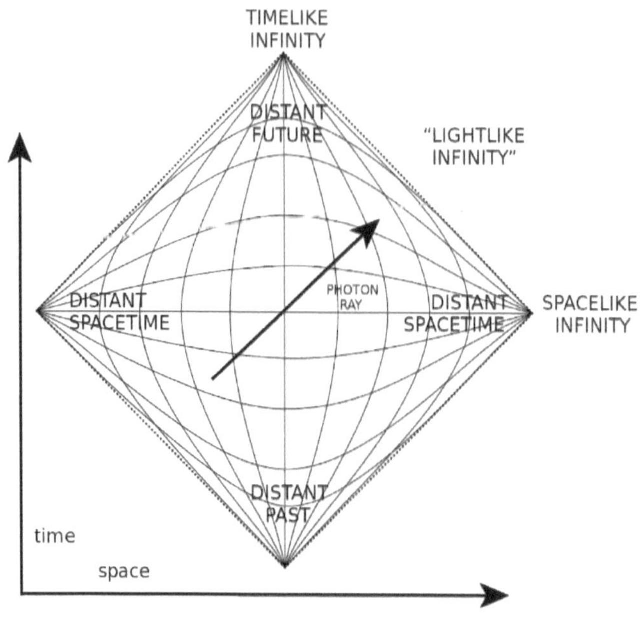

Nothing in this universe can travel faster than the speed of light. In our new coordinate transformed diagram equivalently, our world line (our trajectory in spacetime diagram) can never bend more than 45 degrees. As that would be the same as moving faster than or at the speed of light. This creates a limited freedom (although very hard to even go near) for our worldline to bend. As our universe is four-dimensional, the three-dimensional version of that boundary can be shown below:

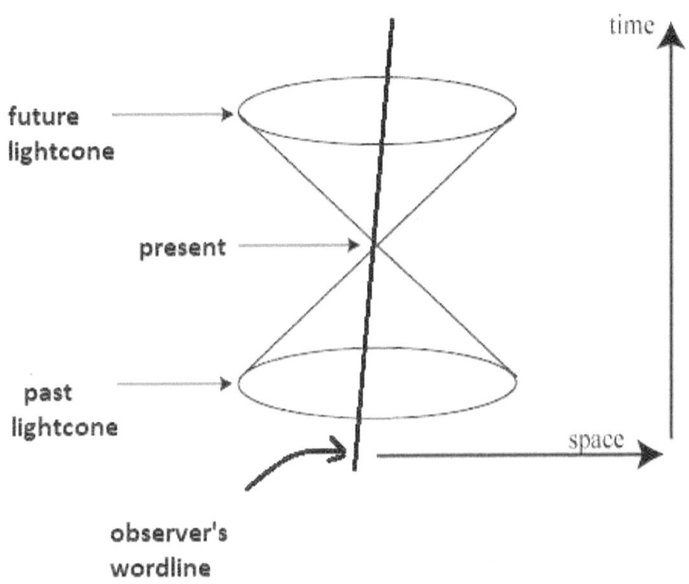

This lightcone is the diagrammatic representation of your Lorentz frame, which you carry wherever you go.

Now imagine a star, 4 light years from you goes into a supernova. This information reaches you in the form of light after four yes and then you realize that a supernova has occurred.

For any event that happened in past and to be witnessed by you, you would need its information to lie inside your past lightcone. Now, imagine you accelerate in some arbitrary direction, this changes the trajectory of your worldline, depending on how you accelerate. Along with your world line, your lightcone also gets tilted (as it's your Lorentz frame). If you keep this acceleration until the time the light from the supernova reaches you, the light from the supernova (which is moving at 45 degrees) could then not be able to match the surface of the light cone (as it's been tilted due to acceleration). So that even remain unevidenced by you. This is just the one event I am talking about, in reality, there would so many events that are not reaching you, due to your tilted lightcone surfaces. So by accelerating, you are tilting your lightcone surfaces and by doing that you are making many events in this universe remain uncommunicated with you. This is equivalent to creating a horizon around yourself. This horizon is called the 'Rindler Horizon'.

We require an ex-extreme amount of energy to boost our rockets to bring a significant tilt in our lightcone. But yes, even a slight tilt can make starlight unreachable to you. So as soon as you stop accelerating and come back to rest or an inertial frame, don't get shocked by looking at the events you never evidenced before.

THE BEAUTY OF BLACK HOLES.

Humans have started studying astronomy by looking up at the sky. Back then we used to find something unique in the sky and then search for it with our telescopes and then study it. After frequent studies, we got to know how this system works (at least a basic algorithm). Laterwards we got tools like never before to study this universe. And after that instead of just looking up, we also started to predict the other possible interstellar objects. A black hole was one of these predictions. We predicted every part of these objects so well that we didn't miss every detail of it. And now we have a real image of a black hole :

The above image is of the M87 Black hole, which belongs to the M87 galaxy. And it's the first real image of a black hole that was released in 2019 and it still astounds me whenever I look at it. The next image shows the image of a Sagittarius A* black hole which belongs to our Milkyway.

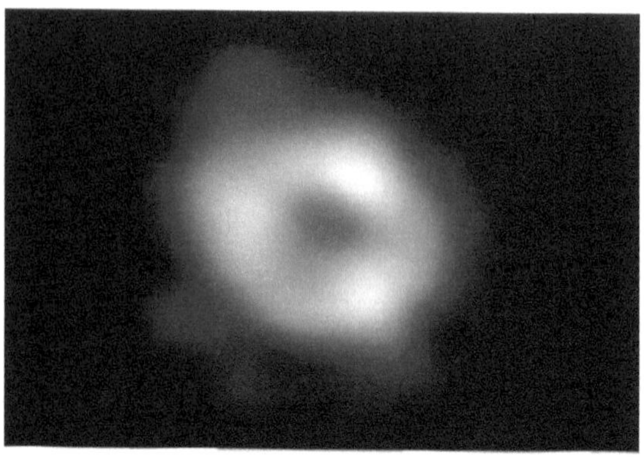

What this image represents is the breakthrough through the fabric of reality. And yes, it is real! We, humans, have successfully predicted something that our naked eyes can't see and finally proved our predictions right by taking a picture of a black hole.

INTERESTING SINGULARITY

Anything that goes into the singularity, ends up being nothing. But, isn't that making the black hole voyage boring? In this chapter, we shall see how can this voyage be made to continue to some other universe than ending up being something at the singularity.

The idea is to connect the singularities of two black holes to each other, creating a tunnel between two universes. This bridge is called a wormhole or Einstein Rosen bridge. It looks like this:

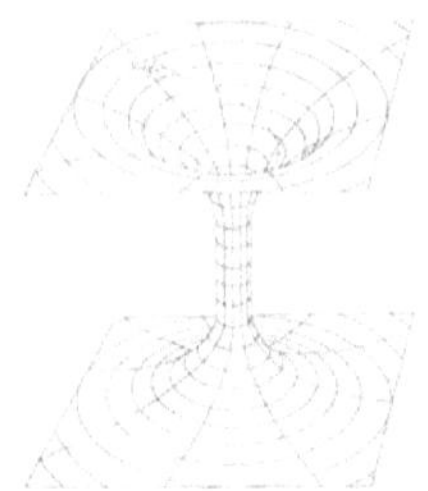

This bridge can also be used to join two universes with opposite running times (physics can be the same or very different in other some other universe!).In that case, a black hole from our universe and a white hole from another would join their singularities. In the universe that is this traveling award in time, the black hole would show the opposite effects than what it shows in our universe. So, in such a universe nothing can enter a black hole, even light can't reach its horizon, and not even information makes it's its inside. We call this object 'A White Hole'. Again, we don't have any evidence for it to exist.

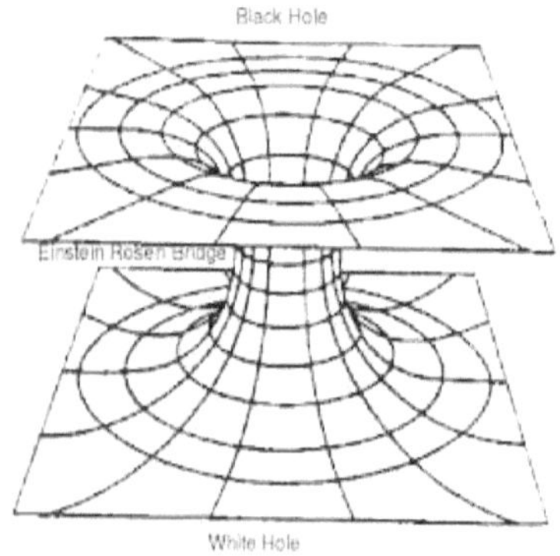

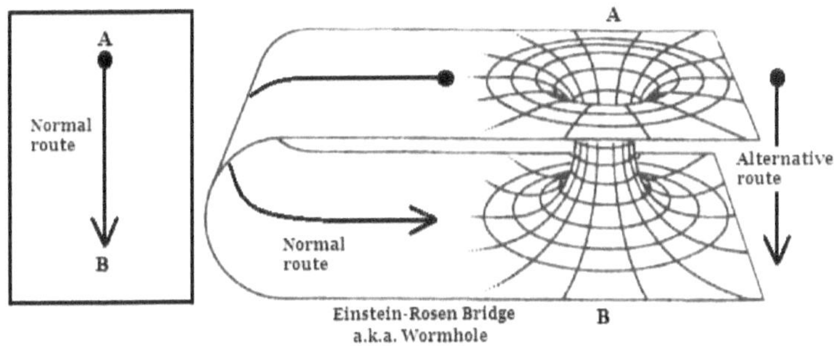

These wormholes can not only join two universes but also two points in our universe alone, creating a shortcut between two points in spacetime. Since a wormhole is a tunnel between two points in 4-dimensional spacetime, it not only can connect two points in space but also time! Creating a time machine for us.

TRAVELING THROUGH TIME

Time is like space, they just have different natures. A wormhole is a bridge between two points in spacetime thus not only can connect two points in 'space' but also 'time' and both of them!

Imagine you are moving at a speed near the speed of light. And he schedules the next meeting after a period of four years. Then at the end of the journey, he will conclude the travel to be as less as 1 minute for him. Returning to the world 4 years later from the time he began his space tourism just would be a future time travel for him. Yes, when you move at relativistic speeds, time slows down for you, and after you return home you discover that you have time traveled in the future.

Now imagine taking one of the mouths of the wormhole to the future by accelerating it at relativistic speeds by our space shuttle. Let's say we go ahead by four years in time, taking one of the mouths of the wormhole along with us in the future. Now as the mouth of the wormhole is always connected to the other mouth of it, it can act as a tunnel between which joins a time gap of four years.

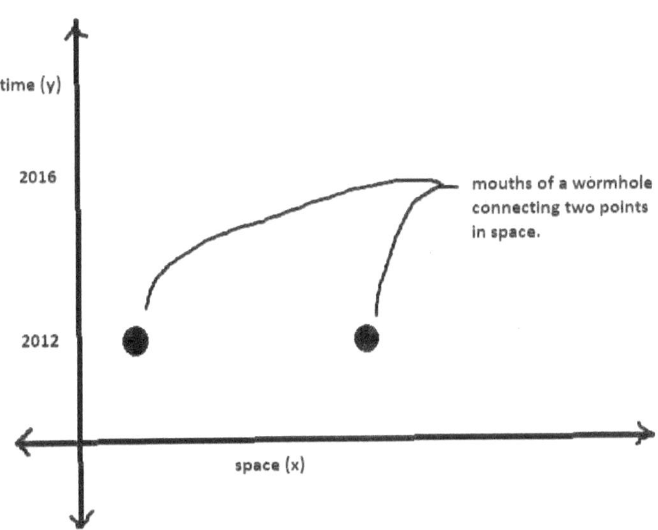

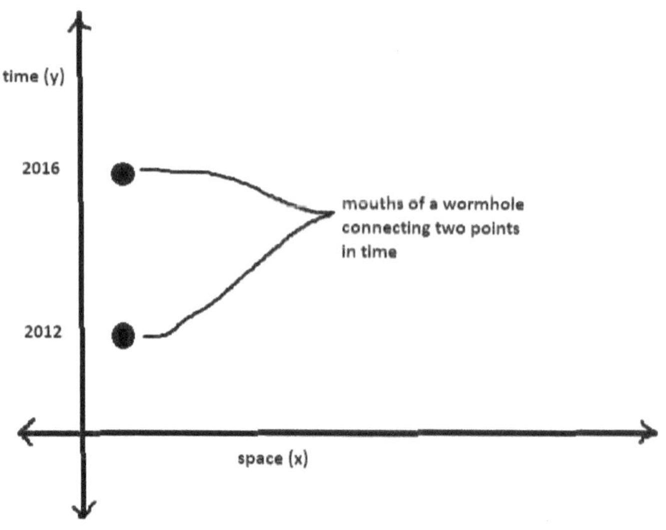

time (y)
2016
2012
mouths of a wormhole
connecting two points
in time
space (x)

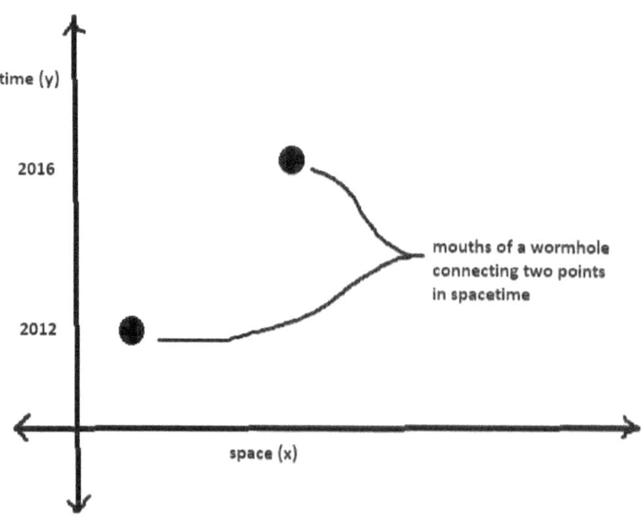

time (y)
2016
2012
mouths of a wormhole
connecting two points
in spacetime
space (x)

So wormholes can connect two points in
space, time and spacetime.

RECIPE OF A LAB-MADE WORMHOLE.

Black hole-black hole and black hole-white hole wormholes aren't complicated to make, even though they might already have been created in our universe. But it is, not easy stuff to create a wormhole in the lab. Spacetime curvature is very common in the universe as its carried by everything in this universe. We treat spacetime curvature due to mass as a positive spacetime curvature. All you need to create a wormhole in the lab is the positive and negative spacetime curvature. Positive spacetime curvature is created by mass. What do you think is needed to create a negative spacetime curvature? A negative mass! Yes, I know that sounds strange. But I will explain, how to create a negative mass-energy density in this chapter.

If you are hearing the term negative mass for the first time then please don't confuse it with antimatter (antimatter has a positive value mass). We are surrounded by an abundance of positive mass around ourselves, the negative mass is very concept. Although we have a possible solution to create it. Let's see how we can do it.

We define negative as something less than zero. So when we say negative mass-energy density, we mean mass-energy density lesser than zero energy mass density or nothing! But we know, that this universe has a layer of energy attached to it, which we call a quantum foam. Which is the zero-level energy of the universe and also what defines 'flat spacetime' and vice versa. Lesser than this energy is what we define as negative energy!

Tale two non-conducting metal plates in a complete vacuum and place them parallel to each other. We know that every mass is accompanied by a wave (quantum mechanics!). All around and in between the metal plates, we have quantum mass-energy. The more the distance between plates, the more the probability of different energy levels being created between the gap. When we have a wide distance between the two plates, the energy around a plate and between the plate differs by almost nothing. But as we get the plates closer to each other, they create a limit for mass-energy to be created inside the gap between the plates. This happens as only smaller mass-energy waves can be built to the small gap, decreasing the variety of the waves that could be formed. The closer we move the plates, the more we put the constraints on the variety of thees that could be formed. At an extremely small gap, only a limited amount of mass-energy can be created inside. Looking at the surrounding, we have two plates surrounded by the point energy everywhere around it, except in between, which is lesser. And what is lesser than zero point energy? The negative energy! Which is our missing ingredient to create a wormhole.

Negative mass repels normal mass and it attracts other negative masses like normal mass. This negative mass would be also useful for us to travel faster than the speed of light (will be discussed later in this book).

LENSING SPACETIME.

Look at the image below. It's the image captured by the jamJamesbWebblescope. Look at the irregular shapes of the galaxy over there. Is that a problem in Webb's image processing? No, and neither in the galaxy nor in the image. The image of a galaxy is just being disturbed by the lens in between. Not a lens made out of glass, it's a lens of gravity!

Light always follows a straight path, but when the entire fabric of spacetime defining its existence is curving, then it has to bend. This bending of light is negligible for smaller masses as they possess lesser spacetime curvature. For masa s as bigger as a black hole, the light just can follow a straight path at all.

This lens of gravity is what we call the gravitational lens. This lens makes black holes detectable to us. So we look at this universe and if some sudden irregularity in the shape of galaxies or star clusters is found, we do research and find a black hole so many times.

Gravitational lenses can also be used to increase the vision of our telescopes. Again, they can help us see the other side of the interstellar object which is blocking the light behind.

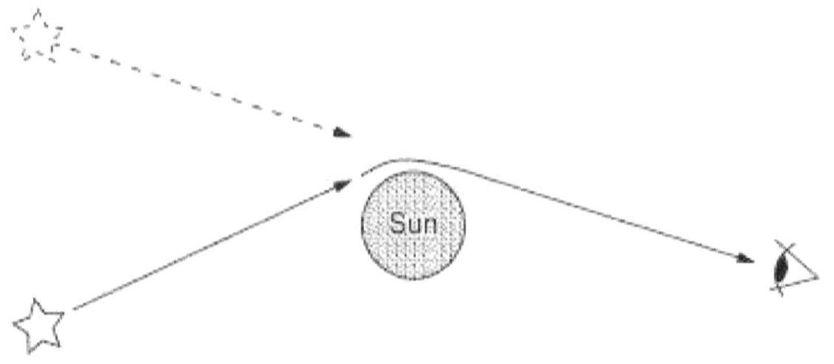

This gravitational lens which lens the light, can lens the gravity. Yes! But to proceed with that let's get some more knowledge about gravity.

WIGGLES IN THE FABRIC OF SPACETIME

From the start, I have made the reader imagine the 4-dimensional fabric of spacetime as a rubber sheet. We can get wiggles in the sheet of rubber, so can we in the fabric of spacetime? Yes! These waves are what we call Gravitational waves. And yes, these are the waves through the fabric of reality.

These waves are created by cataclysmic events like the merging of two neutron stars, black holes, etc. The merging of these massive bodies isn't just like a head-on collision. As these objects possess extreme spacetime curvatures, when they get near collision, they both first revolve around the center of mass of the combined system. This revolving of two massive bodies creates wiggles in the fabric of spacetime. These waves then are sent to the universe outside at the speed of light.

These waves can be detected by the setup, we already have built on earth. And we already have detected many gravitational waves coming from all around the universe. These gravitational waves tell a lot of information about the source they have formed from. Want to know how we detect these waves?

Whatever gravitational waves pass through, have a stretching effect on that object. That's because one side of the body momentarily stays under the trough and the other at the peak of the gravitational wave, thus producing a tidal force on that object. This same stretch is what we use to detect the gravitational wave. But it's not simple as may one think. These gravitational waves cant be detected at one's home. For that,t we have a setup called 'LIGO' in Louisiana, USA.

At LIGO (The Laser Interferometer Gravitational-Wave Observatory), we have an L-shaped system. The arms of this L-shaped system are attached with laser light which meets at a particular point and destructively interferes with each other. When a gravitational wave passes through the arms of the L-shaped system, the arms oscillate. When this happens, the laser light fails to destructively interfere. And flashes of light are detected. These gravitational waves get us a lot of data from their source.

These gravitational waves even pass through us as well. But, we still are not noticed in the normal world or bring any significant difference. But why? That's because gravitational waves are huge and continuous and our local scale just constitutes a flat part of the gravitational wave. So whenever it passes through us, we barely feel any peak or trough of a wave. At the scales of our planet, these waves might produce measurable stress. In short, the larger the body is, the more tidal force it suffers.

Gravitational waves generally that we encounter are smooth and continuous and are pretty safe. But have you ever imagined how dangerous a non-continuous gravitational wave would be? Such a wave can produce the spaghettification effects like a black hole, wherever it goes. Although, we don't know how such waves can be produced.

Gravitational waves aren't forever. They do lose their potential by doing some positive work. Do you wonder where the gravitational waves get their energy from? From nowhere, but from the mass of the source itself! So, the mass-energy from the colliding neutron star or a black hole from somewhere far in the universe is changing the lengths of the L-shaped arms at the LIGO observatory.

It's beautiful to see, how gravitational waves are carrying the mass energy of their source to the distant universe. Gravitational waves are a new way of transmitting energy.

LENSING GRAVITY BY GRAVITATIONAL LENS

Light can be lensed by the optical lens, so can gravity be lensed by the gravitational lens? Yes of course! But in this case, this is 4-dimensional lensing, now let's see its effects.

Gravity can be radiated like light and gravitational lenses can do the work of lensing these waves. Imagine a continuous gravitational wave coming toward the earth. And we have a strong gravitational lens in between like a very massive star which can deflect gravity waves and can lens it on a single point. When we lens sunlight with a simple microscope at the point we get an intensified image of the sun. In the case of light, the simple microscope does the work of collecting the light which falls on its surface and then intensifying it at a point or concentrating it. Here the light that falls on the simple microscope constructively interferes with each other (being a wave), forming an intensified image of the sun. Gravity waves being a wave, show wave-like properties of constructive and destructive interference as well.

So as soon as these gravitational waves interact with our gravity lens, they get lensed to an area if not a point (say). This area is as dangerous as near a black hole. Anyone who gets in this area would have to face the intensified effects of the gravitational waves and have to face the extreme tidal force. This system of gravity waves along with gravitational lenses is no less than a weapon (literally!). Let's say in a war with aliens, we can use this system as a weapon to spaghettified the energy (amusingly), although we don't know how to get the gravitational waves deliberately.

The simple microscope represents an electromagnetic version of a sun on a sheet of paper. Let's say we take our lens near a black hole, then would that create a virtual black hole with no mass? No! Even if we take our lens near a black hole, it would be just a lens a small part of it, and not an entire black hole. And then yes, this part can be confined to a point or diverged or just propagate to some area as it is. This would then create the same spacetime condition over there to the point we wish to project it on.

But yes, if a gravitational wave constitutes complete information about a black hole, and say we lens a tiny part of it, then yes a spacetime impression of the black hole can be created. Instead of impression, it would be better if I call it the black hole itself, as there is nothing more real than spacetime.

SINGULARITIES WITHOUT A BLACK HOLE.

Gravitational radiation or gravitational waves carry a part of the mass of the system they are emitted from. Imagine a hypothetical black hole suddenly starts decreasing its mass in form of gravitational waves until it has no mass. What are we left with then? Spacetime breakdown! JusttJust that! Have we ever seen anything like these In our universe? Not yet! These are called Naked Singularities.

Currently, according to our observation, every singularity is being covered by a black hole around naked singularity would just be then an area or point that cuts spacetime and is not covered by a black hole. But why can't see the naked singularities? The answer is very simple, the information inside a black hole does not make it along the surface of our lightcone.

But why is every singularity covered by an event horizon? From the spacetime curvature around the singularity to the event horizon of the black hole, the spacetime is drastically curved as compared to the universe in the neighborhood. There is so much spacetime curvature, that it marks a strong horizon boundary outside. And the lightcones outside the horizon just don't match up with the ones inside the horizon which is in the fascination of the horizon).

So can an observer inside even horizon see the naked singularity? Yes, It's,s then possible.

SPACETIME INSIDE A BLACK HOLE.

Spacetime flows inside a black hole like water in the washbasin, moving towards the spacetime breakoff (singularity). When you cross the event horizon of the black hole, time and space will switch their roles. And our spacetime becomes time-spacee'. Under the influence oftime-spacee, you would move in space just in one direction like time and there is no other option. That's because space has become time-like, and as 'proper time' never stops flowing. Inside a black hole, you can't resist the flow of your one-dimensional space.

Now, what about the time? The time inside a black hole becomes space-like. So, now here time shows spacelike behavior, and one starts looking at time as one can do in space in the normal world. So one always has the freedom to look at the past, present, and future.

However one might like this 'time-space', one can enjoy it only for a short period. This is because of the continuous flow of time-space inside a singularity, where time and space have a breakoff.

THE TIMESPACE UNIVERSE

The fabric of reality or the fabric of spacetime inside an event horizon does not get destroyed, but does not even stay as 'spacetime'! It gets completely flipped. Our laws of physics do not work at the singularity due to the absence of spacetime. The same thing even happens inside an event horizon as spacetime is time-space and our laws of physics although are just made for spacetime.

Do you want an example? Ok! According to the fourth law of thermodynamics, the temperature can never be equal to 0 K and there will be always some temperature associated at the tiniest scales of spacetime. This temperature is due to the quantum foam. This law fails inside a black hole. Shortly, this is because originally the law says that the 0 K is unattainable in spacetime and not in time-space. We define energy as the ability of a system to do some work. The Quantum energy is nothing but the continuous production of virtual particles. So when we locate any event in spacetime, there is always some production of tiny virtual particles. These virtual particles but needs a plane to come into the existence and annihilate each other. But as space becomes time is inside a black hole. And as this timelike space just flows in the forward direction and one dimension, there is no possibility of the reaction of virtual particles as that needs two dimensions. So the absolute zero temperature can be attended in time-space if not in spacetime (our universe)

HOW TO CORRECTLY IMAGINE SPACETIME?

We, humans, love to know new things, we love to explore, and we ask questions about everything. It's our nature to wonder about things and ask questions. We find answers the answer to these questions and get our curiosity satisfied to know something new and bizarre. We are on the planet earth and we have the farthest place any human has gone until now is the moon, but our questions have reached the edge of the cosmos beyond it when nobody knows whether the universe has an end?.

It's our habit that we first believe in real things or the things which our sensory organs can see, feel or touch. When it comes to spacetime, we do see the space around us and we can feel time. Many people don't even believe time as a real thing, they will simply tell you that time is a clock (which is nuts!). Even if we believe in space and time, both as real entities, we forget about their connection. Space and time are not separate things are connected to each other, forming 'spacetime'. Now even if I say spacetime one might think about space around us and feel the running time.

It's hard to visualize spacetime, in fact impossible, as spacetime is four-dimensional and humans can see 4 dimensions. Now I want a reader to suppress one dimension from the three dimensions of space and imagine spacetime as a dynamic evolution of these two-dimensional space sheets with respect to time. Now since we are looking at 3-dimensional spacetime. It might be easier for a reader to understand it correctly.

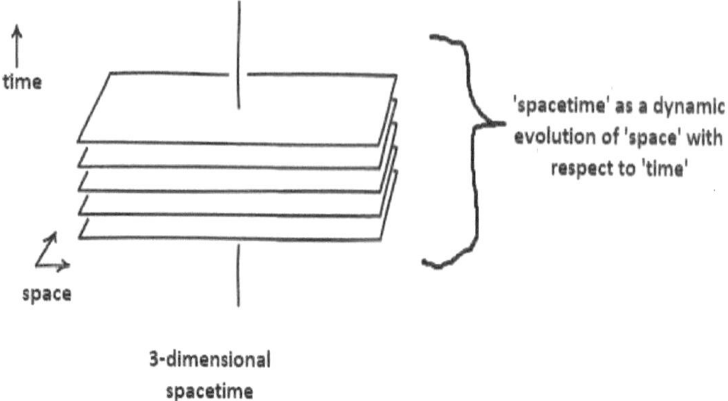

Now feel the real-world 4-dimensional spacetime the same way. Look around the 3-dimensional space around you and convince yourself that whatever you looking at is not the present but the past. Since light takes time to reach from everything to reach you, everywhere you look, every sound you here is just an event in past. So wherever you point your fingertip, it's not a point in space or time but rather spacetime, you are pointing to an 'event'! And by doing our ongoing imaginations we are heading toward the future Our present is the sign we are here, the past is what we can see all around us and the future is what we are heading towards and we always will. What we are perceiving is the 4-dimensional spacetime. This spacetime is nothing but our universe itself. Let's study more about it from now onwards.

WHERE IS OUR UNIVERSE LOCATED?

We are located in our universe which is completely made of a spacetime continuum. But did you ever wonder where it's located?

Our universe is made of 4-dimensional spacetime, three of which are spatial and another one is the time dimension. What about extra dimensions? Yes! Our universe is located in the higher dimensions. We also call it hyperspace. We, humans, can just perceive up to 4 dimensions and not more than that, so hyperspace is not possible for us to imagine.

The force of gravity even needs the extra dimensions to exist. Because gravity is the curvature in 4-dimensional spacetime and when you curve a particular dimension, you take it to the next dimension. Is there any other universe in hyperspace? Let's discuss that in the chapter ahead in this book. For now, let's note that the universe exists in the extra dimensions or hyperspace.

HOW LARGE IS OUR UNIVERSE?

Where does the universe end? or How large is our universe? We, humans, believe in discreteness, we believe that everything has a starting and an end, and we believe that everything has a limit. When this thought is applied to our universe, where does our universe end? Nlet'sets say a show you its end, then the next question would be what's beyond that end boundary? And so on. So don't you think our question about how large is our universe does not make any sense, that is because there is no sensible answer to the question.

Our universe is made of spacetime and from nothing called 'space and time'. So our question which mentions just 'space' and not time would be completely wrong. We should instead ask where-when our universe will end.

First let's see, where our universe or spacetime is in hyperspace. Nowhere in the hyperspace (extra dimensions) I am safe to ask the question, where our universe is located? As we are locating spacetime in the extra dimensions and not in the same domain where spacetime has its existence. So in that sense, our spacetime starts from the big bang and ends in a big crunch which is the events that ever occurred and going to occur in our universe. Let's study them in detail in the next chapter.

BIRTH OF OUR UNIVERSE

The birth of our universe is the birth of spacetime. And the entire universe started its life from an infinitely dense point called a singularity. Spacetime ceases to exist at such points. But such points are even the birth of the fabric of reality that makes us real.

The birth of spacetime is the birth of energy. We describe the big bang as an expansion of the spacetime continuum or our universe from an infinitely small point. After the big bang, there was a sudden increase in the expansion rate of the universe which we call the inflation period. After which our universe is continuously expanding.

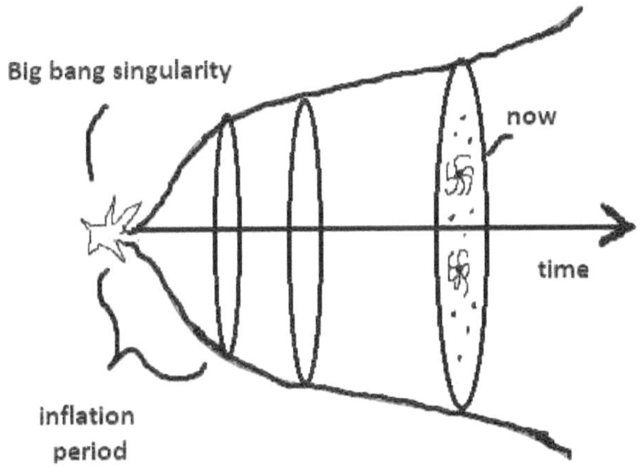

The image above is the 3-dimensional representation of the 4-dimensional big bang.

The singularity which gave rise to our universe is called the Big bang singularity or the cosmic singularity, as at that time the entire cosmos was inside it. Many might misunderstand the expansion of the universe as a stretching of the fabric of spacetime, but that's not the case. As our universe is expanding, it's creating new spacetime in itself. So spacetime is not getting stretched, but rather gets regenerated inside.

The origin of the universe is the origin of energy. So the origin of energy dates back to the origin of our universe., spacetime curvature or the force of gravity came into the existence. So among all fundamental forces, the force of gravity was the first to come. Next to gravity came the strong force and then the electromagnetic force and in last, the weak force. Between the time period of the big bang and the creation of the force of gravity, all the fundamental forces of the universe were in the form of one super force combined. That's amazing, right?

The big bang singularity and the singularity of a black hole are very similar. The difference is that the big bang singularity is the creation of spacetime, while is the end of spacetime. Again, the big bang singularity spews out the spacetime and black hole singularity swallows it down. There is so much similarity in both of them that the opposite of a black hole would be a near-to-perfect replica of the big bang singularity. The opposite of a black hole is a white hole. Wait! Is our universe inside a white hole? Let's go ahead In the next chapter and discuss that.

ARE WE INSIDE A WHITE HOLE?

The big bang structure of our universe is very much similar to the opposite of a black hole. So are we inside a white hole? The answer is no! There is no problem for us to be inside a white hole. But our universe isn't. That's the fact.

But how did we conclude that? White holes are the time reverse version of the black holes, they do everything opposite two the black holes. So white hole has the same spacetime metric that a black hole has. Our universe is well described by FLRW metric locally. This FLRW metric even fits well inside a part of the Shwarzchild metric of the black hole. So if our universe is inside a black hole, it would be harder for us to know whether we are inside a black hole.

But still no! Because according to CMB (The cosmic microwave background - The map of our universe), the metric of our universe at large scales doesn't commute with the metric of a black hole, and there is a huge difference. Also, the evolution of spacetime from the cosmic singularity is very different from the evolution of spacetime from the white hole singularity.

This explanation sometimes reminds me of the Huygens principle. This says that every slit, when coming into contact with light, can be thought of as a new source of light. In our case, every white hole in any universe can be thought of as the one that has given the rise to the universe itself. That's interesting. Although keep in mind that our universe is not inside a white hole.

THE EXPANSION SPEED OF THE UNIVERSE.

Our universe is expanding since the day it formed. Currently, it's expanding at the rate approximately described by the Hubble law. Which says that farther is the any object from you in space, the more it will speed up from you (Due to cosmic expansion). This law has no exception for the speed of light. So anything that according to the law is moving faster than the speed of light is then moving faster than light. Wait, but that's a violation of the laws of physics, as nothing can travel faster than the speed of light. No! physics is still not violated here.

The laws of physics are just made for the constituents of the universe and the universe itself can violate them. Yes, it might surprise the reader, but yes it's that way. During the cosmic expansion, the universe is expanding at the speed of light and not its constituents. So it's not a violation of the laws of physics. So the speed at which the universe expands can at any speed, there is no violation of physics here.

Does the expansion of the universe have any effect on the spacetime curvature of the earth or any interstellar object? The answer is no! The expansion of the universe is only significant at very wide distances. At a local scale, the force due to expanding universe would be very very less than the counterbalancing electromagnetic and other forces, even gravity. So, it might just take anything even at the scale of the galaxy, along with it, but will not make it stretch. So, gravity or spacetime curvature or any astronomical object doesn't get stretched due to the expansion of the universe.

Our observable universe is roughly 93 billion lights apart in length but provided that the age of our universe is roughly 14 billion years old, we need the part of the universe that far to be traveling more than the speed of light, which is what can be explained by the expanding structure of our universe.

Technically, light from the edges of the observable universe should reach us after 41.5 billion light years, but still, it makes its way to us. That's because, even if the object is taken away from us by the expanding universe, but no so the light it emitted. Instead of going along with it, the light gets stretched This stretching of light is called redshift as the original light due to this stretching loses its frequency. That is why these lights appear in the low energy range to us.

HOW UNIVERSE ENDS?

The end of our universe is the end of our universe! So, to find the end of the universe, we should search for an event, rather than seeking for the spatial boundary of the universe (as our universe is spacetime and not 'space and time).

The universe is going to end in a similar event to the big bang, but just with an opposite functioning. We call this event a Big crunch. So, before any such event happens, we have the continuous and then drastic contraction of our universe. Contraction to what? A singularity! Which would be the same as a black hole singularity, where everything is crushed down to a singularity. The Big crunch singularity would not bring any stress over the spacetime during its contraction. This is because, spacetime does not really shrink in size, but rather starts to disappear getting smaller and smaller. We live in a big crunch era where the universe has just started contracting, would barely feel this contraction as it's only significant at an extremely large scale. But eventually, as it shrinks further and the temperature of the universe increases, the things we see in the sky with our naked eyes as well as we would start getting affected by it. To escape such a cataclysm there is no choice than traveling to some other universe to escape a big crunch. Ahead in this book, I will show our progress in this universe and how we always overcome the cosmic threats we might possess in the future. For now, let's study our significance in this universe.

PHYSICIST AND PHYSICIST PLUS PHILOSOPHER

According to classical physics, we and every living thing no matter what do not differ at all! But the fabric of spacetime on which the entire physics is dependent gives us a high priority as well as significance. I am not saying that physics is wrong. Physics just doesn't study the things like these, because physics is not made for that. That's why I say that a physicist plus a philosopher understands this universe much more better than the physicist alone. A physicist mostly studies and sticks to one field of physics and gives opinions, but a physicist plus philosopher will always see different laws of physics to explain our universe and everything!

A physicist always sticks to mathematics and the laws of physics but does not understand that every basic law of physics is formed by our observation and experience. And there is something that we can't experience directly, but via some agency and it would not be fine to blame that agency for the particular phenomenon or event we might experience. Every physicist plus philosopher gives himself the freedom to see that part which lies before that agency. A physicist plus philosopher will always see this universe with a lens that does not violate the laws of physics and makes us a significant part of this universe. So a physicist studies real things, but a physicist plus philosopher studies reality! So every physicist should also study philosophy.

But why do we talk about our significance in this universe? Because cosmology is the study of the city, its study not only explains the constituents of this universe but also how and why spacetime ever formed and all related to its formation. To deal with such topics even the field of physics like quantum mechanics relies on observers or us a lot!

So let's study the relationship between our existence in this universe in detail. But before that, we would again need a bit of quantum theory.

THE WAVE AND PARTICLES.

Quantum physics is the subject that predicts the behavior of sub-atomic particles, and which cant be described by any other field of physics. Quantum mechanics predicts the motion of particles with quite well accuracy if not approximately, this less precision or uncertainty in the subject of quantum mechanics, makes it a language of probability. According to quantum physics, every particle is a wave and vice versa.

A light ray when passed through a slit, does show the interference pattern on the screen, showing the wavelike behavior. But when the same light when shun on a metal for the photoelectric effect, then light behaves like a particle. Not only light but this is seen to be happening with subatomic particles that do have a little mass as well. For example, in a nuclear reaction, an alpha particle does not possess much energy to penetrate through a potential barrier wall it has due to its electromagnetic interaction due to the rest of the subatomic particles in the radioactive atom, but it still makes it through its barrier. This phenomenon can only be explained by the wave behavior of the alpha particle. The alpha particle treated as a wave, and the probability for it to escape the potential barrier was always non-zero as a part of the wavefunction was always lies beyond the potential barrier. This makes alpha particle to transfer their existence to the other side of the barrier and again behave like a particle.

The higher the speed of anything and the more massive it is, the less wavelike behavior anything posses. We call this a de Broglie wavelength which related the particle's wavelength with its mass and velocity. Since subatomic particles are very light in weight, they have long wavelengths associated with them. Normal-day objects are also made of subatomic particles, but they are connected with electromagnetic energy. Treated every particle let's say of your couch, they do possess significant wavelength at their scale. So that means that your couch is made of waves. But each of these is different from the other, spanning in either direction, making your couch as a whole become more difficult but not impossible to be treated as a wave. This applies to every normal day objects and larger ones.

THE OBSERVER AND THE UNIVERSE.

Gravity is the force that makes gigantic bodies less wavelike and electromagnetic force is the force that makes normal day objects less wavelike. There is no problem for even a normal day object to behave like a wave. But they don't because of the irregularity in the wavefunctions of the atoms they are made of. So gravity and electromagnetic force are not allowing for the subatomic particles in any normal-day object or larger to behave wavelike entirely, if not individually. This makes the planets, stars, moon, oceans, etc behave like a particle. Without the particle nature of matter, life just wouldn't have existed. But the same wavelike nature of light heats our planet and makes alpha decay possible thing.

The universe its fundamental forces, the laws of physic the nature of matter, the fundamental constants, and many more parameters are well arranged at high precision in this universe to make life on our planet a possible thing.

The universe needs observers to exist. Many of our experiments in quantum physics prove that the observation crystallizes the wavefunction or uncertainty into reality. This is the observer effect. The observation in the observer effect is just not confined by our direct or indirect measurement, but by our existence in this universe in general. So the observation that observers exist has reached up till the past until the big bang to crystallize the laws of physics that would make life a possible thing. Even a slight change of any of the fundamental constants say at a 10theh decimal place is yet enough to make life not a possible thing in this universe. The universe or spacetime since its formation from the big bang is very much connected to the observer now and in the future. Such a universe is called a participatory universe. And the universe highly resembles it. See the image below:

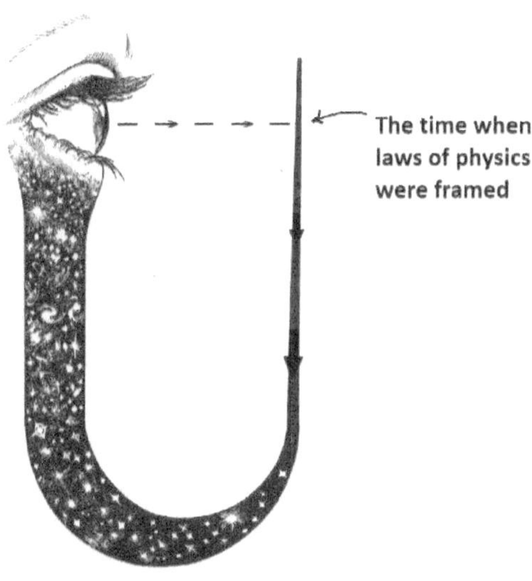

The above image is the description of the participatory universe and the shape of the diagram has nothing to do with the shape of our universe, it's just a short description of the entire content discussed above. The participatory universe says that the participants of this universe (we) have influenced this universe from its very start by making the observation now sitting on the earth.

All fine, that we exist and make the universe to form in a way that would allow our existence. But what if the universe chooses to die at the end of the day and why would our universe ever do that? Let's go ahead with that.

THE LAWS OF THERMODYNAMICS.

The laws of thermodynamics are very basic laws of nature. That's because they rely on symmetry, and this universe wants symmetry. Everything in this universe tends to attend to the low energy state and move further to do that. And the laws of thermodynamics -. To date, nothing has violated these laws. There are 4 laws in thermodynamics in total, lets's understand their meaning to understand the fate of our universe further.

The first law of thermodynamics :

'IF TWO THERMODYNAMIC BODIES ARE EACH IN THE THERMODYNAMIC EQUILIBRIUM WITH THE THIRD BODY THEN ALL THREE BODIES ARE IN EQUILIBRIUM WITH EACH OTHER.'

This law is pretty obvious as it gives the idea of normal world heat transformation. And yes this law is what applies to every system of this universe. The transfer of heat is the process of becoming cold and to become cold is achieving the low energy state, which is what our universe is like and tends towards.

The second law of thermodynamics :

'THE TOTAL ENERGY IN OUR UNIVERSE REMAINS THE CONSTANT'

The birth of the universe is the birth of energy. As a finite universe, the energy in it is also finite. And every real thing in this universe is made out of this energy. The sun gives this energy to the earth, the plants on earth absorb this energy, herbivore feed on these plants so on. So this energy can be transformed from one form to the other, but never be created.

This energy also includes quantum energy which lies everywhere in the universe at the Planck scale. This energy is stored in various forms. The form that can keep huge energy more than any other form Is mass. In fact, mass is not a form of energy, it is the same as energy. So in general we restate the second law of thermodynamics as follows :

'THE TOTAL MASS-ENERGY IN OUR UNIVERSE REMAINS THE CONSTANT'

Which is the very basic law in physics.

The third law of thermodynamics:

'THE ENTROPY OF THE UNIVERSE ALWAYS INCREASES

Entropy is the measure of the dispersal of energy through any process. It's a parameter we always take into account for any engineering project. Let's say an airplane is boarding at some speed. The fuel of an airplane just is not making the airplane move, but the airplane is making a lot of sounds. This decreases the efficiency of fuel. The sound of the airplane is somehow thus contributing to the entropy created by the airplane engine.

Entropy is the 'information'. Our universe is the stock of events and everything that happens here is only events, these events are contributing continuously to the total entropy of the universe. This second law of thermodynamics tells us that everything in this universe keeps on creating information. Aging is the very basic form of information creation that contributes to the total entropy of the universe. So everything in this universe does age. So entropy has its meaning attached to time. So until the moves in the future direction, the entropy of the universe will keep on increasing.

Third law of thermodynamics:

'THE ABSOLUTE ZERO TEMPERATURE IS UNATTAINABLE'

Yes! There is a limit to how much a system can be cooled down. An absolute zero kelvin or - 273.15 °C is that limit. Temperature is the measure of the movement of gas. In other words, this law states that a complete 100% percent vacuum is not possible. The entire universe is covered by quantum foam, so there is always some energy available, even in the empty space.

THE FATE OF OUR UNIVERSE

We, humans, have made the universe attain the exact laws of physics that are necessary to allow our existence. But what if the entire universe chooses to end itself? Yes! the laws of thermodynamics give us the hint universe is going to end sometime in the future But what we humans can do to still ensure our future in the cosmic eternity when the entire universe decides to perish itself?

The first law of thermodynamics says that the total energy in this universe remains constant. This energy is stored in various forms in nature. The very basic form of fuel we use is the food we eat, which is we directly or indirectly get from plants, we get proteins from them, which are then converted into energy. The source of this energy is the sun as the plants use solar energy to prepare their food via photosynthesis. Later after that, we started using energy stored in the form of coal. It takes lakhs of years for the remains of animals and plants to form coal. We are still using the coal as our energy need in many sectors of industry. The energy in the coal has come from the energy of plants or animals that coal has formed from in fact have their energy is gained through plants which gain their energy from the sunlight. Due to science and technology, we are at a point where we can convert mass into energy and use that energy to do some work. Almost a century ago we were using all the forms of energy that the sun makes available to us. But later after then we discovered that mass is the same as energy, and we learned how to use the mass to run our turbines and make destructive bombs that would threaten humanity. The energy we get from the sun is even created from the mass-energy inside the sun itself. So all the energy available to us is in the form of solar energy which we have been using for decades and even now has the mass of the sun as its primary source. It's the energy required for life to evolve has been achieved from the mass of the sun itself.

But, this story is not going to go on forever. The second law of thermodynamics says entropy of the universe always increases. This entropy is the reason why everything ages and ultimately going to die out. Our sun continuously fusion its lighter elements into heavier ones. This fusion reaction will go on until we reach the iron. Iron being not able to fuse, our sun will

Simply form a white dwarf which will again convert itself into a black dwarf and nothing else. Sun almost covers every energy need of us. And since it's not permanent as everything else in this universe, how are we going to shine our lamps? But still for our sun to die out we have several billion years in our hands for us to reach some other energy source.

But again the problem is that, as the sun fuses into the heavier elements, these elements are then sent into the outer layers of the sun. So our sun is continuously increasing its volume. So we just can't chill by looking at the sun to form a black dwarf from our sky. The sun soon (next 4 to five billion years later) is going to come and fall on us. I mean the sun is going to get enough big that its atmosphere will completely swallow the earth. And as the sun is coming closer to us, more would be an increase in heat energy, so we have to leave earth much earlier. So we have a limited time to search for our next energy source In the next 2 billion years, humans would have developed enough technology that we can convert a mass of an asteroid completely into energy (say). But what is more convenient than a ready-made energy form? So, we definitely are going to look forward to visiting and getting settled around some other star. Yes! The same star that one can see in the night sky.

After harnessing the energy from around every star in this universe, we would look for harnessing the energy from the entire galaxy. After harnessing energy from every star, every galaxy, we would now seek energy from the last candidates which can stay alive till the cosmic eternity. The black holes would be the last energy source available source in this universe. They can provide us with an immense amount of energy. So in eternity, the black hole would be our last energy source.

I guess we have gone too ahead in the future, we not even thought about, that the universe is ever going to end. The big crunch is the suspected end of our universe. So in eternity, let's see we get to know that our universe is shrinking its size by seeing the recession of the galaxies at the cosmic scale. Then what? We have to hurry up! We have to somehow again perform the tradition of securing our existence. But now what's next when the entire universe is dying? Ok, then let's travel to some other universe! Using the wormholes. The technology in eternity would be built enough for us to create a 5-dimensional between the four-dimensional universes. And even we can find some readymade black holes serving this purpose for us.

But wait! traveling to some other universe is just not as easy as one might think. As we need the laws of physics, including every fundamental constant over there to be the same as in our universe with a high precision. And provided we could find one, we going to do the entire mentioned process again and so on. So, we (the observers) are always going to secure our existence whatever happens till the eternity and in some other universe.

THE MANY WORLDS.

Quantum mechanics is the language of probability, it explains the dynamics at the sub-atomic level in terms of probability. But does even matter at larger scales like at normal world dimensions? There is a famous experiment in quantum mechanics. We call it the Schrodinger's cat experiment.

In this experiment, we have a box in which we have a live cat, a glass vessel containing poisonous gas, and a nuclear reactor with a ball projectile. The nuclear in the setup if successfully does some work, then it converts some of the nuclear energy into mechanical energy which is then transferred to the ball inside. This ball is set to hit the glass vessel containing the poison, which if broken can kill the cat inside the box.

An observer closes the box. The nuclear reactor in the setup only reacts when there is a successful alpha particle emission. So indirectly, if an alpha particle gets emitted, it kills a cat! The chances of the emission of the alpha particle are given in the form of a wave by quantum mechanics, and so does the fate of the cat!

So, whether the cat is dead or alive now is completely described by the wave described by quantum mechanics. If one asks the observer whether the cat is dead or alive, he would then say both, the cat is dead and at the same time alive! Because that event is a wave now, and we as a wave is spread, there is no reason to treat a particular part of it specially and say that cat is dead or alive. So the fate of the cat is now a wave. But when the observer opens the box, the observer collapses the wave function into one result whether a cat is dead or alive.

According to quantum mechanics, many worlds interpretation says that a part of the wavefunction of the fate of the cat has made its way to some other universe! It says that if a cat is dead in our universe then it's alive in some other. So the observation in either universe collapses the wavefunction to a single result. If this applied to normal world things, then you might have not even been born in one universe but in another, you are!

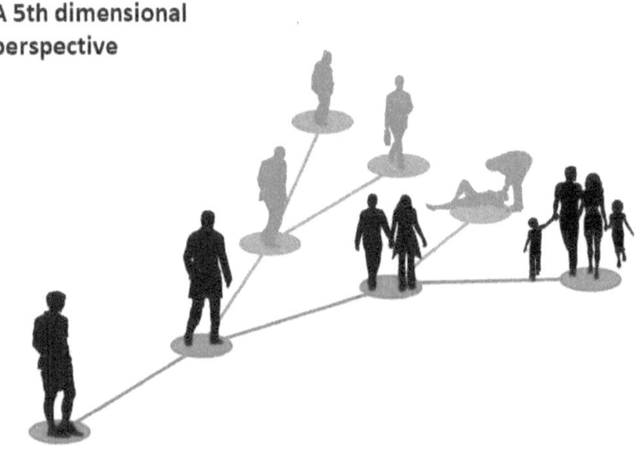

An electron wave when passed through a slit and caught by the observed detector, produces an interference pattern. Similarly, the the wavefunction of man above in the image is producing the interference pattern in different universes.

Our universe is one of these many worlds and we form a multiverse in the hyperspace. The number of universes in the multiverse can be infinite in number. Currently, we don't have any reason to limit it to a particular constant.

TIME TRAVEL IN THE PAST.

One can travel in time in the future along with his Lorentz frame but not in the past. In fact no one can do that in our universe. Our universe or spacetime is made that way, that the time component attached to space always moves in the future. So, it's ok to travel in the future, but not in past (strictly!). It's not even possible by a wormhole.

When you accelerate, time slows down for you. Time can slow down till it ultimately stops, which would happen if you travel at the speed of light. And if you travel still faster then you travel backward in time. But in reality that does not work like that. Firstly, achieving the speed of light is not possible in the real world,d and secondly traveling backward in time is still impossible, The spacetime becomes a time-space inside a black hole and it separates itself from what our universe is. Thus some laws of physics break over the same Same applies when you travel backward in time. Our spacetime is

made in a way that time always means a dimension that runs forward and traveling backward in time would simply mean spacetime where time moves backward and that's not the definition of spacetime and our laws of physics like inside the event horizon just break down over there. So in general we can say that spacetime can never run backward in time.

Now, what about that-made wormholes? If a wormhole is made then the only people who can go into the future are the people the from past and not the opposite. Because even inside the wormhole the time in spacetime is directed toward the future. So until we have a wormhole made from a white hole and a black hole where spacetime has pulled on either side, time travel in past is not possible.

In our universe, we have yet to see negative mass and again a white hole made out of it is yet to be detected.

LIGHT AND ITS SPEED.

What makes light so special that nothing can travel at its speed? Ok, to be honest, there is nothing special in light. The specialty is in the speed it travels at. From the start, we have been calling it the speed of light. But that speed limit has nothing to do with light. The actual value of this speed limit is 299 792 458 m / s. We call this the speed of causality.

The speed of causality marks the boundary between causal relations or cause and effect. For example, in our light case , any cause that happened in past can affect us unless it lies inside our past lightcone. And similarly, any effect can't have any cause for us until it lies inside our future lightcone. The boundaries of the lightcone are margined by nothing but a speed limit. A speed limit that nothing in this universe can cross. This speed limit is what we call the speed of causality.

The speed of causality is the universal speed limit of any given Lorentz frame. When anything tries to reach this speed, the more massive it becomes. And more is the energy we require to push anything to the speed of light.

Since light and graviton are massless particles, they always travel at this speed.

THE LIGHT'S PERSPECTIVE.

The faster we move, the slower the time runs for us. Since light travels at the speed of causality, it almost stops elapsing for it. Imagine an observer looking at the sun, the photons from the sun in his perspective take 8 minutes to reach his eyes. But from the photon's perspective,e it made it into the observer's eye since the time it was emitted (since time doesn't elapse in the photon's perspective). So the question is is it predestined for a photon to ever go into the observer's eye? The answer is yes, but only predestined in the photon's perspective! Let's learn why.

The below image shows the lightcone of a particle having some mass and in the inertial frame (at rest)

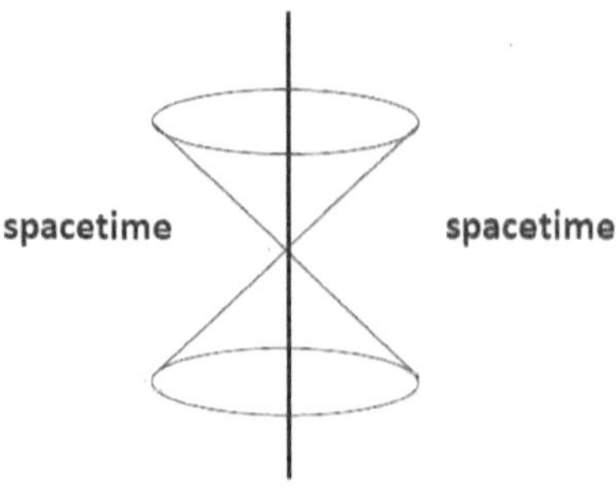

As the particle accelerates its lightcone gets tilted (remember, as we discussed). Let's say the lightcone tilts in the counterclockwise direction. The only permissible tilt any particle can go is the tilt of 45 degrees as after that tilting of lightcone is the same as moving faster than the speed of causality which is against the laws of physics. Light travels at the speed of causality, below image shows its lightcone.

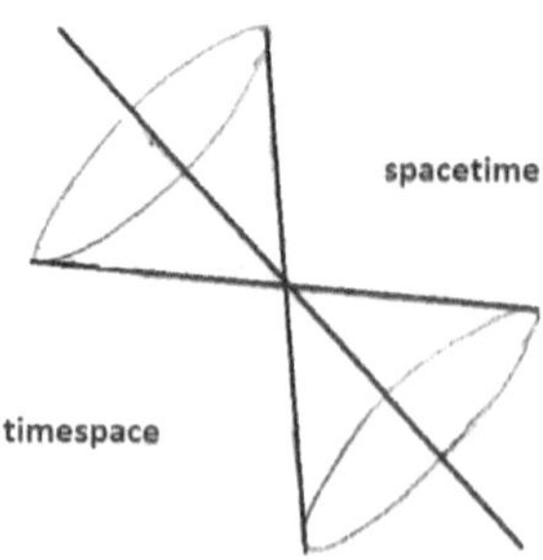

Light or photon, traveling at the speed of light always has its lightcone being 45 degrees tilted.

And when spacetime gets tilted more than 45 degrees, it switches time to space and space to time in our Penrose diagram on which we define the light cone. In the above image of the lightcone of a photon, one can see that 50% part of the future and past lightcone enters the timespace and the other stays in spacetime.

As we live in spacetime, we can just see the spacetime behavior of light. But by studying the timespace + spacetime nature of light, many characteristics of light can be explained. For example why light only moves in one dimension? That's because, in time-space space becomes timelike, so there is only one direction to move, while light moves in one dimension it can move in any direction as we see it, this is just due to the spacetime freedom of light.

To answer the main question we addressed at the beginning of this chapter, it predestined in light's perception to enter into the observer's eye. This is because a part of the light is in timespace, where time is spacelike. So always has here the freedom to see in time as we do in space, looking at the past, feeling the present, and seeing the future.

I always say that anything different than spacetime is not our universe as the laws of physics are just best fit for spacetime and nothing more or less than that. Provided that one of the two parts of the lightcone lies in the timespace, light or photon can be entirely be thought of as a particle that exists in our universe and simultaneously not and we are just witnessing its real part (spacetime part) that exists in our universe.

OUR UNIVERSE DURING PLANCK ERA

The four fundamental forces of our universe the electromagnetic, gravitational, Weak Force, and Strong Force based on which we now see our universe the way it is. These forces were not separate as they now are but rather combined into a single super force. The existence of the super force dates back to the time (Planck time) when our universe was just $10-43$ seconds old before which the only dominating force was the super force, and that's it. After the Planck era (which is the period between the big bang and Planck time), the super force started disintegrating into gravitational, strong, electromagnetic, and weak forces. The gravitational force was the force that disintegrated first after which then the strong force and after that the electromagnetic and weak forces.

Can we get this super force into the laboratory?

Well, provided we have the condition hot enough we can do that. All thing you have to do is to get to the Planck temperature (**10^{32} Kelvin**). If you go hotter than

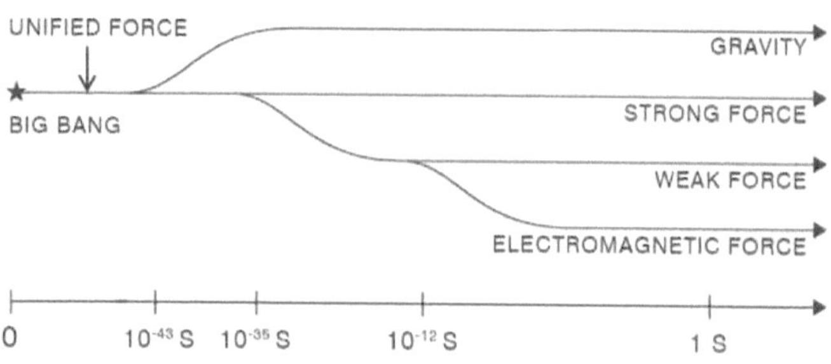

that then all the four fundamental forces start unifying into a single super force,

Newton's theory of gravity works well in some if not all the cases to explain our universe and what takes the handle of the whole thing is now the

The general theory of relativity that till fails at the singularity and thus gives the handle to the quantum gravity over there. Imagine an equation that describes the dynamics of the entire universe and is smaller enough to write on a part of the paper. This universal theory explains everything from the very dawn of creation to the big crunch and everything in our universe. This universal theory is the **theory of everything**. The Grand Unified Theory is the unification of strong, weak, and electromagnetic forces into say a single force not comprising gravity. The act of describing the unification of all the fundamental forces of our universe (super force) in physics by the way of mathematics is what will describe the conditions and dynamics of our universe during the Planck era. The theory is still not complete as we have no complete theory of gravity. Except that we have unified the other three fundamental forces. Many of our observations prove Einstein's General Relativity right, except at the quantum scale and the singularity conditions. Shortly, the acceleration of gravity is due to the spacetime curvature according to general relativity and at the quantum scale and the singularity conditions, quantum gravity is what works. The unification of general relativity and quantum gravity is the key. Quantum gravity as a theory has to get confirmed via experiments. In LHC (Large Hydron Collider) when we collide two particles at relativistic speeds and say in the soup of new particles created after the disintegration we find a graviton. Such outcomes would prove quantum gravity true. Up to date, there is no evidence of finding a graviton yet.

The best theory we know that attempts to unify the Grand Unified theory and the universal theory of gravity which will get us to the Theory Of Everything are the String Theory.

We shall not discuss string theory in this book. This chapter was just for a bit of knowledge. Before we move toward the end of this book, let's go along with some more interesting chapters I would like an astrophysics newbie to know.

.

THE FORCE OF GRAVITY AND OTHER FUNDAMENTAL FORCES

When we say force, it reminds us of an acceleration brought in mass. But we don't always have to push that mass by ourselves. In the case of the fundamental forces of the universe, this force seems very mystical. I broke the mystery of gravity by explaining to the reader what exactly it is and how it works. Every fundamental force in this universe needs a field to operate. The field responsible for weak force is the weak field, for the electromagnetic force, It's the electromagnetic field, and for strong force, it's the gluon field.

Force the force of gravity the field is 4-dimensional spacetime itself! So the force of gravity is a very basic thing to have. Since the fabric of spacetime is the fabric of reality, the fields of all fundamental forces have their meaning attached to the field that is responsible for the force of gravity. This is how the force of gravity Is different from other fundamental forces o this universe.

THE FLOW OF TIME AND GRAVITY.

Let's say you are near a black hole, your future has yet not been governed by the horizon completely. Have you ever thought about why we are moving toward a black hole.?Even if you know that you are in rest. Of course, it's a gravity. But what is pulling you into in the first place? It's the time! The nature of time in spacetime is to flow and when it flows into some event you have a possibility to meet that event. If your time finds itself in the black hole singularity, that possibility turns into definiteness and you inevitably move towards that event which has now become an event in your time than just space.

So, there is no gravitational acceleration without the moving nature of time. And the only way we can control the flowing nature of spacetime is by using our boosters. It's amazing to know this as we whenever look at the clocks, we just think about whether we would get late for our business meeting or something else, but we never think of time as a dimension that gives rise to gravitational acceleration.

THE CENTRE OF OUR UNIVERSE.

Where is the center of the universe? Again, this is the wrong question. We are again mentioning just 'space' in our question and we don't even are thinking about time and forget about the combination of both 'spacetime'! We have to do that, whenever we imagine things at the cosmic scales. A failure of imagining our universe as a spacetime continuum would simply give us the wrong answers.

When we locate anything in spacetime, we actually locate an event. An event that lies between a big bang and a big crunch. So our question should be what event is the center of the universe? So it can be any event that is in the center of spacetime. It can even be an event of you reading this book or a second ago or after that.

THE QUANTUM ZENO EFFECT

According to quantum mechanics, observation collapses the wavefunction. Or unless a particle is seen as a particle, it was always a wave. So quantum mechanics has changed our view to look at the world around us. Imagine a wave function of an event, and imagine an observer observing it. The moment an observer looks at the wavefunction, it turns itself into the real world. And the moment he sees another side, the real world becomes the world of wavefunctions. This makes wave function to evolve when not observed and collapse when observed

The quantum Zeno paradox is an idea that talks about the continuous observation of a wave function. Which makes a wave function hard to evolve. So continuous observation just stops the wavefunction from evolving. Again, for example,

Imagine that we provide some energy to an electron at which it gets to some excited state, and soon as that transition from a lower energy level to a higher energy level takes place we keep providing it energy (unaltered). Then an electron will never hit the ground state again at least until we shut our laser beams off. The mere act of providing an electron the energy is the process of observation or measurement which makes its way to a higher energy state. An unaltered series of measurements makes the system remain unperturbed. The example given here was a very basic example of the quantum Zeno effect

Zeno's paradox in a sentence: Repeated observation of a system halts its progress.

ALIENS

To this date, there is no solid evidence of the existence of aliens. But still, is there any one of them out there? Any planet in another system in the habitable zone (a zone suitable for the creation of life) gives us the chance of having extraterrestrial life. Our planet earth, which is in the solar system, is in the milky way that has more than 3200 star systems in it, and still, there are more than three trillion estimated galaxies in the observable universe. Thus in that sense, it's highly probable for the alien lives to exist. Can Extraterrestrial life if exist be more intelligent than us? well! we don't know. Aliens may even be type 2 civilizations at the level of which technology is enough advanced to harness the complete power of any star. Aliens also may be advanced human beings. If they meet us we would have a great opportunity to learn from them. Also, they can be bacteria or any unicellular animal too.

Jupiter's moon Europa is quite likely to sustain life in it as there remains the liquid water underneath its ice sheets which have widths of several kilometers wide. Europa gets sunlight but not that bright to maintain the water in a liquid state beneath So from where does Europa get that energy to maintain the liquid water beneath the ice sheets? the answer is the tidal gravity of Jupiter. The tidal gravity or tidal force of Jupiter brings out some deformation in the shape of Europa, this mechanical energy is then converted into heat energy by breaking various molecular bonds providing some heat energy source to maintain the liquid water under the ice sheets of Europa. One also can compare this with the tennis ball when it gets hit by a racket. When the tennis ball is hit by a tennis racket, it gets deformed and in fact, gets heated due to the mechanical deformation acted on it. In the same way, Europa receives its heat by which it maintains liquid water beneath the ice sheets on it. But here what is deforming Europa is the tidal gravity of Jupiter. This deformation is not as great as in the tennis ball analogy but quite enough to provide energy to the planet to keep water in the liquid form beneath the ice sheets on it. A mission is planned by NASA till 2024 to study still in detail to investigate the suitable conditions for the existence of life on Europa and if we get some positive results there, then it would truly be one of our greatest achievements of us.

Well! what about the aliens from some other universe? Great idea right? If the multiverse exists, then we have a greater chance of life harboring in each of the individual universes in it. So if we are alone in 'our universe ' then there might be aliens existing in another universe too. What about the bulk beings? (Multidimensional beings) There is no proof of the existence of bulk beings. If God exists in 10 or 11 dimensions we may call him a bulk being or bulk alien or whatever you like. There is a famous equation of aliens that gives the probability of the existence of extraterrestrial life in our universe. It considers each and every parameter like the number of planets that are habitable, the fraction of stars orbited by planets, etc. This equation is also known as the Drake Equation.

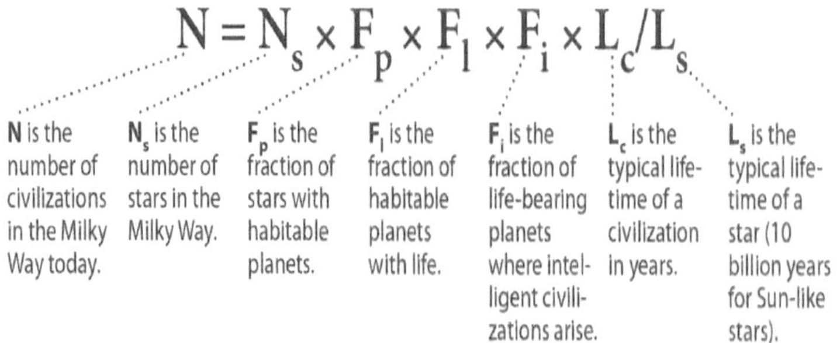

$$N = N_s \times F_p \times F_l \times F_i \times L_c / L_s$$

N is the number of civilizations in the Milky Way today.

N_s is the number of stars in the Milky Way.

F_p is the fraction of stars with habitable planets.

F_l is the fraction of habitable planets with life.

F_i is the fraction of life-bearing planets where intelligent civilizations arise.

L_c is the typical life-time of a civilization in years.

L_s is the typical life-time of a star (10 billion years for Sun-like stars).

The presence of us in this universe as the only life in existence in actuality is a rare case. If we are in reality all alone in our universe, could it be interpreted as there is some relation between the existence of humanity and the existence of the universe? Roughly speaking it could be!

Even some observations of cosmic microwave background like the axis of evil suggest us that we humans have some special role in this universe! But still, this is speculation and there is no solid basis proving it strongly. But until we investigate each and every suspected habitable zone for life, our hunger for searching for extraterrestrial life will persist.

THE CONCLUSION

Until now the reader has learned so many things about this universe. These outcomes come from our observation at our scale. In fact, this is how we tried to explain things. We see things around us and we try to apply them to a larger scale and whether it works the same way. And when it doesn't we get a flaw in our theory or we seek for a new theory that will explain the part of the universe where our laws fail and the ones we see in our neighborhood. For example,e we think of a collision of two balls heading toward each other as a head-on collision. And when we apply the same notion to astronomical spherical bodies like planets, stars, neutron stars, or anything massive, they just don't appear to make a head-on collision. They rather revolve around their common center of mass first and then collide. So this was never predicted by Newtonian gravity. And we found a flaw in the laws of physics backthen found general relativity to explain this behavior. This is how we keep changing the way we look at this universe.

Throughout the book, you may have noticed that I am not using the word 'universe' alone, but rather using 'this universe' every time I mention it. This is because I am afraid that the reader will misunderstand the concept. Because the laws of physics can be different in some other universes we can't apply our concepts to some other universe! So I always explain the concepts of 'Our universe' or 'This universe'. I want a reader to understand this.

I skipped mathematics in this book and my book 'ASTROPHYSICS FOR NON-MATHEMATICIANS'. I know many might hate this subject or can't find it easy. But that's just because they had got the wrong teacher. The same problem even I had until I met a teacher that changed my perspective to learn mathematics. It's the language that helps physics so much to encode the information and to predict the results. We started maths by counting things on our fingers, and now we have them in form of equations that we use to encode the information In physics. These equations gave us so much information to us. Without these equations, we could never have figured out that the energy is same as mass. With the help of these equations, we can explain the sub-atomic world, we predict the motions of the stars without going near to them. So maths is really interesting if one is under the right mentor or one studies it by himself.

I have always tended to explain cosmology to everyone, no matter what field is anyone from. That's because I think that this subject is for everyone. That's because whatever we see around us, we think of it as our universLet'sets say a super extrovert man who cares a lot about society and its belief system, barely has a time to give himself the to look up in the sky and gaze at the stars. For such people, the society they live in is their universe. For a housewife, her family is her universe. For an excellent employee, impressing his boss is his universe. This is what most people do. But when we give a part of our time to ourselves to explore the real universe, we learn a lot.

A stressed, depressed person or an extrovert takes society so much for granted that he never knows what the real universe is. When you gaze at the stars, everything in the human universe seems irrational. And we start taking society lightly. This makes us believe in ourselves than listen to society and thrive in the shifting world. Thus looking at the real universe is what everyone should do. Thus cosmology at a basic level and star gazing should always be part of everyone's daily life. I am doing it since the age of 5 and I still wonder about our cosmos.